Philippe Halsman

## *About the Author*

Pierre Teilhard de Chardin (1881–1955) was born in France and ordained a Jesuit priest in 1911. Trained as a paleontologist, Teilhard did research at Musée National d'Histoire Naturelle in Paris and fieldwork in China, where in 1929 he codiscovered the celebrated "Peking Man" fossils. In his writings, he sought to reconcile his spiritual and scientific beliefs, producing a vision of man as evolving toward the divine. His unorthodox theological positions were at odds with Catholic doctrine and led to a strained relationship with Jesuit leaders, who forbade him from publishing his writings. *The Phenomenon of Man* became a bestseller when it was posthumously published in France in 1955.

Sir Julian Huxley (1887–1975) was one of the twentieth century's leading evolutionary biologists. Among his numerous distinctions, Huxley was the first director general of the United Nations Educational, Scientific, and Cultural Organization (UNESCO) and cofounder of the World Wildlife Fund.

*Pierre Teilhard De Chardin*

# *THE PHENOMENON OF MAN*

WITH AN INTRODUCTION
BY SIR JULIAN HUXLEY

HARPERPERENNIAL MODERNTHOUGHT

NEW YORK • LONDON • TORONTO • SYDNEY • NEW DELHI • AUCKLAND

HARPERPERENNIAL MODERNTHOUGHT

Originally published in French as *Le Phénomene Humain*, copyright © 1955 by Editions du Seuil, Paris.

First published in English in 1959 by Wm. Collins Sons & Co., Ltd., London, and Harper & Row, Publishers, Incorporated, New York. A revised translation was first published in 1965.



HarperCollins books may be purchased for educational, business, or sales promotional use. For information, please e-mail the Special Markets Department at SPsales@harpercollins.com.

First Harper Colophon edition published 1975. Reprinted in Perennial 2002.

First Harper Perennial Modern Thought edition published 2008.

Library of Congress Cataloging-in-Publication Data is available upon request.

ISBN 978-0-06-163265-5 (Harper Perennial Modern Thought edition)

22 LSC 20 19 18 17

# Contents

## BOOK TWO: LIFE

CONTENTS

# TRANSLATOR'S NOTE

Perhaps a word may be permitted about some of the lesser problems involved in the translation of this book.

The author's style is all his own. In some instances he coins words to express his thought—'hominisation', for instance, or 'noosphere'—and in others he adapts words to his own ends, as when he talks of the 'within' and the 'without' of things. His meaning, however, should become apparent as his thought unfolds, and I have dispensed with cumbrous efforts at defining his terms.

As far as possible I have dispensed with italics for his neologisms—they are repeated too often to stand italicisation in a work already thickly sprinkled with italics for emphasis. I have also, in obedience to the conventions of typography in England, eliminated the author's initial capitals for all abstract nouns such as 'science', 'life', 'thought', and also for 'world', 'universe', 'man' and other such key-words of his work. There were disadvantages in this decision, but at least the printed page looks more normal to the English reader.

A number of people have contributed to the translation, some by substantial paper work, others by suggestions; and the outcome is in a sense a joint effort. Outstanding among participants are Mr. Geoffrey Sainsbury, Dr. A. Tindell Hopwood, Professor D. M. MacKinnon and Mr. Noel Lindsay. At times versions or suggestions have been conflicting and I have had to take it on myself to make an editorial decision. The translators' notes appear in square brackets. I should like to thank my wife, without whom it would have been impossible to produce this version. Finally, I must take on myself responsibility for the inadequacies that still persist.

BERNARD WALL

## *Introduction by Sir Julian Huxley*

*The Phenomenon of Man* is a very remarkable work by a very remarkable human being. Père Teilhard de Chardin was at the same time a Jesuit Father and a distinguished palaeontologist. In *The Phenomenon of Man* he has effected a threefold synthesis—of the material and physical world with the world of mind and spirit ; of the past with the future ; and of variety with unity, the many with the one. He achieves this by examining every fact and every subject of his investigation *sub specie evolutionis*, with reference to its development in time and to its evolutionary position. Conversely, he is able to envisage the whole of knowable reality not as a static mechanism but as a process. In consequence, he is driven to search for human significance in relation to the trends of that enduring and comprehensive process ; the measure of his stature is that he so largely succeeded in the search. I would like to introduce *The Phenomenon of Man* to English readers by attempting a summary of its general thesis, and of what appear to me to be its more important conclusions.

I make no excuse for this personal approach. As I discovered when I first met Père Teilhard in Paris in 1946, he and I were on the same quest, and had been pursuing parallel roads ever since we were young men in our twenties. Thus, to mention a few signposts which I independently found along my road, already in 1913 I had envisaged human evolution and biological evolution as two phases of a single process, but separated by a 'critical point', after which the properties of the evolving material underwent radical change. This thesis I developed years later in my *Uniqueness of Man*, adding that man's evolution was unique in showing the dominance of convergence over divergence : in the same volume I published an essay on *Scientific Humanism* (a close approximation to Père Teilhard's *Neo-Humanism*), in which I independently anticipated the title of Père Teilhard's great book by describing humanity as a phenomenon, to be studied and analysed by scientific methods. Soon after the first World War, in *Essays of a Biologist*, I made my first attempt at defining and evaluating evolutionary progress.

In my Romanes Lecture on *Evolutionary Ethics*, I made an attempt (which I now see was inadequate, but was at least a step in the right direction) to relate the development of moral codes and religions to the general trends of evolution ; in 1942, in my *Evolution, the Modern Synthesis*, I essayed the first comprehensive post-Mendelian analysis of biological evolution as a process : and just before meeting Père Teilhard had written a pamphlet entitled *Unesco : its Purpose and Philosophy*, where I stressed that such a philosophy must be a global, scientific and evolutionary humanism. In this, I was searching to establish an ideological basis for man's further cultural evolution, and to define the position of the individual human personality in the process—a search in which I was later much aided by Père Teilhard's writings, and by our conversations and correspondence.

*The Phenomenon of Man* is certainly the most important of Père Teilhard's published works. Of the rest, some, including the essays in *La Vision du Passé*, reveal earlier developments or later elaborations of his general thought ; while others, like *L'Apparition de l'Homme*, are rather more technical.

Père Teilhard starts from the position that mankind in its totality is a phenomenon to be described and analysed like any other phenomenon : it and all its manifestations, including human history and human values, are proper objects for scientific study.

His second and perhaps most fundamental point is the absolute necessity of adopting an evolutionary point of view. Though for certain limited purposes it may be useful to think of phenomena as isolated statically in time, they are in point of fact never static : they are always processes or parts of processes.

The different branches of science combine to demonstrate that the universe in its entirety must be regarded as one gigantic process, a process of becoming, of attaining new levels of existence and organisation, which can properly be called a genesis or an evolution. For this reason, he uses words like *noogenesis*, to mean the gradual evolution of mind or mental properties, and repeatedly stresses that we should no longer speak of a cosmology but of a *cosmogenesis*. Similarly, he likes to use a pregnant term like *hominisation* to denote the process by which the original protohuman stock became (and is still becoming) more truly human, the process by which potential man realised more and more of his possibilities. Indeed, he extends this evolutionary terminology by employing terms like *ultra-hominisation* to denote the deducible future stage of the process in which man will have so far transcended himself as to demand some new appellation.

With this approach he is rightly and indeed inevitably driven to the conclusion that, since evolutionary phenomena (of course including the phenomenon known as man) are processes, they can never be evaluated or even adequately described solely or mainly in terms of their origins : they must be defined by their direction, their inherent possibilities (including of course also their limitations), and their deducible future trends. He quotes with approval Nietzsche's view that man is unfinished and must be surpassed or completed ; and proceeds to deduce the steps needed for his completion.

Père Teilhard was keenly aware of the importance of vivid and arresting terminology. Thus in 1925 he coined the term *noosphere* to denote the sphere of mind, as opposed to, or rather superposed on, the biosphere or sphere of life, and acting as a transforming agency promoting hominisation (or as I would put it, progressive psychosocial evolution). He may perhaps be criticised for not defining the term more explicitly. By *noosphere* did he intend simply the total pattern of thinking organisms (i.e. human beings) and their activity, including the patterns of their interrelations : or did he intend the special environment of man, the systems of organised thought and its products in which men move and have their being, as fish swim and reproduce in rivers and the sea ?[1] Perhaps it might have been better to restrict *noosphere* to the first-named sense, and to use something like *noosystem* for the second. But certainly *noosphere* is a valuable and thought-provoking word.

He usually uses *convergence* to denote the tendency of mankind, during its evolution, to superpose centripetal on centrifugal trends, so as to prevent centrifugal differentiation from leading to fragmentation, and eventually to incorporate the results of differentiation in an organised and unified pattern. Human convergence was first manifested on the genetic or biological level : after *Homo sapiens* began to differentiate into distinct races (or *subspecies*, in more scientific terminology) migration and intermarriage prevented the pioneers from going further, and led to increasing interbreeding between all human variants. As a result, man is the only successful type which has remained as a single interbreeding group or species, and has not radiated out into a number of biologically separated assemblages (like the birds, with about 8,500 species, or the insects with over half a million).

Cultural differentiation set in later, producing a number of psychosocial units with different cultures. However, these 'interthinking groups', as one writer has called them, are never so sharply separated as are biological species ; and with time, the process known to anthropologists as cultural diffusion, facilitated by migration and improved communications, led to an accelerating counter-process of cultural convergence, and so towards the union of the whole human species into a single interthinking group based on a single self-developing framework of thought (or noosystem).

In parenthesis, Père Teilhard showed himself aware of the

[1] In *Le Phénomène Humain* (p. 201) he refers to the *noosphere* as a new layer or membrane on the earth's surface, a 'thinking layer' superposed on the living layer of the *biosphere* and the lifeless layer of inorganic material, the *lithosphere*. But in his earlier formulation of 1925, in *La Vision du Passé* (p. 92), he calls it 'une sphère de la réflexion, de l'invention consciente, de l'union sentie des âmes'.

danger that this tendency might destroy the valuable results of cultural diversification, and lead to drab uniformity instead of to a rich and potent pattern of variety-in-unity. However, perhaps because he was (rightly) so deeply concerned with establishing a global unification of human awareness as a necessary prerequisite for any real future progress of mankind, and perhaps also because he was by nature and inclination more interested in rational and scientific thought than in the arts, he did not discuss the evolutionary value of cultural variety in any detail, but contented himself by maintaining that East and West are culturally complementary, and that both are needed for the further synthesis and unification of world thought.

Before passing to the full implications of human convergence, I must deal with Père Teilhard's valuable but rather difficult concept of *complexification*. This concept includes, as I understand it, the genesis of increasingly elaborate organisation during cosmogenesis, as manifested in the passage from subatomic units to atoms, from atoms to inorganic and later to organic molecules, thence to the first subcellular living units or self-replicating assemblages of molecules, and then to cells, to multicellular individuals, to cephalised metazoa with brains, to primitive man, and now to civilised societies.

But it involves something more. He speaks of complexification as an all-pervading tendency, involving the universe in all its parts in an *enroulement organique sur soi-même*, or by an alternative metaphor, as a *reploiement sur soi-même*. He thus envisages the world-stuff as being 'rolled up' or 'folded in' upon itself, both locally and in its entirety, and adds that the process is accompanied by an increase of energetic 'tension' in the resultant 'corpuscular' organisations, or individualised constructions of increased organisational complexity. For want of a better English phrase, I shall use *convergent integration* to define the operation of this process of self-complexification.

Père Teilhard also maintains that complexification by convergent integration leads to the intensification of mental subjective activity—in other words to the evolution of progressively more conscious mind. Thus he states that full consciousness (as seen in man) is to be defined as 'the specific effect of organised complexity'. But, he continues, comparative study makes it clear that higher animals have minds of a sort, and evolutionary fact and logic demand that minds should have evolved gradually as well as bodies and that accordingly mind-like (or 'mentoid', to employ a barbarous word that I am driven to coin because of its usefulness) properties must be present throughout the universe. Thus, in any case, we must infer the presence of potential mind in all material systems, by backward extrapolation from the human phase to the biological, and from the biological to the inorganic. And according to Père Teilhard, we must envisage the intensification of mind, the raising of mental potential, as being the necessary consequence of complexification, operating by the convergent integration of increasingly complex units of organisation.

The sweep of his thought goes even further. He seeks to link the evolution of mind with the concept of energy. If I understand him aright, he envisages two forms of energy, or perhaps two modes in which it is manifested—energy in the physicists' sense, measurable or calculable by physical methods, and 'psychic energy' which increases with the complexity of organised units.[1] This view admittedly involves speculation of great intellectual boldness, but the speculation is extrapolated from a massive array of fact, and is disciplined by logic. It is, if you like, visionary: but it is the product of a comprehensive and coherent vision.

It might have been better to say that complexity of a sort is a necessary prerequisite for mental evolution rather than its cause. Some biologists, indeed, would claim that mind is generated solely by the complexification of certain types of organisation, namely brains. However, such logic appears to me narrow. The brain alone is not responsible for mind, even

[1] See, e.g., C. Cuénot, *Pierre Teilhard de Chardin*, Paris, 1958, p. 430. We certainly need some new terms in this field: perhaps *neurergy* and *psychergy* would serve.

though it is a necessary organ for its manifestation. Indeed an isolated brain is a piece of biological nonsense, as meaningless as an isolated human individual. I would prefer to say that mind is generated by or in complex organisations of living matter, capable of receiving information of many qualities or modalities about events both in the outer world and in itself, of synthesising and processing that information in various organised forms, and of utilising it to direct present and future action—in other words, by higher animals with their sense-organs, nerves, brains, and muscles. Perhaps, indeed, organisations of such complexity can only arise in evolution when their construction enables them to incorporate and interiorise varied external information : certainly no non-living, non-sentient organisation has reached anything like this degree of elaboration.

In human or psychosocial evolution, convergence has certainly led to increased complexity. In Père Teilhard's view, the increase of human numbers combined with the improvement of human communications has fused all the parts of the noosphere together, has increased the tension within it, and has caused it to become 'infolded' upon itself, and therefore more highly organised. In the process of convergence and coalescence, what we may metaphorically describe as the psychosocial temperature rises. Mankind as a whole will accordingly achieve more intense, more complex, and more integrated mental activity, which can guide the human species up the path of progress to higher levels of hominisation.

Père Teilhard was a strong visualiser. He saw with his mind's eye that 'the banal fact of the earth's roundness'—the sphericity of man's environment—was bound to cause this intensification of psychosocial activity. In an unlimited environment, man's thought and his resultant psychosocial activity would simply diffuse outwards : it would extend over a greater area, but would remain thinly spread. But when it is confined to spreading out over the surface of a sphere, idea will encounter idea, and the result will be an organised web of thought, a noetic system operating under high tension, a piece of evolutionary machinery capable of generating high psychosocial energy. When I read his discussion of the subject, I visualised this selective web of living thought as the bounding structure of evolving man, marking him off from the rest of the universe and yet facilitating exchange with it : playing the same sort of role in delimiting the human unit of evolution and yet encouraging the complexification of its contents, as does the cell-membrane for the animal cell.

Years later, when at the University of California in 1952, this same vivid imagination led Père Teilhard to draw a parallel between the cyclotron generating immense intensities of physical energy in the inwardly accelerating spiral orbits of its fields of force, and the entire noosphere with its fields of thought curved round upon themselves to generate new levels of 'psychical energy'.[1] How his imagination would have kindled at the sight of the circular torus of Zeta, within whose bounding curves are generated the highest physical energies ever produced by man !

Père Teilhard, extrapolating from the past into the future, envisaged the process of human convergence as tending to a final state,[2] which he called 'point *Omega*', as opposed to the *Alpha* of elementary material particles and their energies. If I understand him aright, he considers that two factors are co-operating to promote this further complexification of the noosphere. One is the increase of knowledge about the universe at large, from the galaxies and stars to human societies and individuals. The other is the increase of psychosocial pressure on the surface of our planet. The result of the one is that the noosphere incorporates ever more facts of the cosmos, including the facts of its general direction and its trends in time, so as to become more

[1] *En regardant un cyclotron* : in *Recherches et débats*, Paris, April 1953, p.123.

[2] Presumably, in designating this state as Omega, he believed that it was a truly final condition. It might have been better to think of it merely as a novel state or mode of organization, beyond which the human imagination cannot at present pierce, though perhaps the strange facts of extra-sensory perception unearthed by the infant science of parapsychology may give us a clue to a possible more ultimate state.

truly a microcosm, which (like all incorporated knowledge) is both a mirror and a directive agency. The result of the other is the increased unification and the increased intensity of the system of human thought. The combined result, according to Père Teilhard, will be the attainment of point Omega, where the noosphere will be intensely unified and will have achieved a ' hyperpersonal ' organisation.

Here his thought is not fully clear to me. Sometimes he seems to equate this future hyperpersonal psychosocial organisation with an emergent Divinity : at one place, for instance, he speaks of the trend as a *Christogenesis* ; and elsewhere he appears not to be guarding himself sufficiently against the dangers of personifying the non-personal elements of reality. Sometimes, too, he seems to envisage as desirable the merging of individual human variety in this new unity. Though many scientists may, as I do, find it impossible to follow him all the way in his gallant attempt to reconcile the supernatural elements in Christianity with the facts and implications of evolution, this in no way detracts from the positive value of his naturalistic general approach.

In any case the concept of a hyperpersonal mode of organisation sprang from Père Teilhard's conviction of the supreme importance of personality. A developed human being, as he rightly pointed out, is not merely a more highly individualised individual. He has crossed the threshold of self-consciousness to a new mode of thought, and as a result has achieved some degree of conscious integration—integration of the self with the outer world of men and nature, integration of the separate elements of the self with each other. He is a person, an organism which has transcended individuality in personality. This attainment of personality was an essential element in man's past and present evolutionary success : accordingly its fuller achievement must be an essential aim for his evolutionary future.

This belief in the pre-eminent importance of the personality in the scheme of things was for him a matter of faith, bu of faith supported by rational inquiry and scientific knowledge It prevented him from diluting his concept of the divine principle inherent in reality, in a vague and meaningless pantheism, just as his apprehension of the entire process of reality as a system of interrelations, and of mankind as actively participating in that process, saved him from losing his way in the deserts of individualism and existentialism.

He realised that the appearance of human personality was the culmination of two major evolutionary trends—the trend towards more extreme individuation, and that towards more extensive interrelation and co-operation : persons are individuals who transcend their merely organic individuality in conscious participation.

His understanding of the method by which organisms become first individualised and then personalised gave him a number of valuable insights. Basically, the process depends on cephalisation —the differentiation of a head as the dominant guiding region of the body, forwardly directed, and containing the main sense-organs providing information about the outer world and also the main organ of co-ordination or brain.

With his genius for fruitful analogy, he points out that the process of evolution on earth is itself now in the process of becoming cephalised. Before the appearance of man, life consisted of a vast array of separate branches, linked only by an unorganised pattern of ecological interaction. The incipient development of mankind into a single psychosocial unit, with a single noosystem or common pool of thought, is providing the evolutionary process with the rudiments of a head. It remains for our descendants to organise this global noosystem more adequately, so as to enable mankind to understand the process of evolution on earth more fully and to direct it more adequately.

I had independently expressed something of the same sort, by saying that in modern scientific man, evolution was at last becoming conscious of itself—a phrase which I found delighted Père Teilhard. His formulation, however, is more profound and more seminal : it implies that we should consider interthinking humanity as a new type of organism, whose destiny it is to realise new possibilities for evolving life on this planet.

Accordingly, we should endeavour to equip it with the mechanisms necessary for the proper fulfilment of its task—the psychosocial equivalents of sense-organs, effector organs, and a co-ordinating central nervous system with dominant brain ; and our aim should be the gradual personalisation of the human unit of evolution—its conversion, on the new level of co-operative interthinking, into the equivalent of a person.

Once he had grasped and faced the fact of man as an evolutionary phenomenon, the way was open towards a new and comprehensive system of thought. It remained to draw the fullest conclusions from this central concept of man as the spearhead of evolution on earth, and to follow out the implications of this approach in as many fields as possible. The biologist may perhaps consider that in *The Phenomenon of Man* he paid insufficient attention to genetics and the possibilities and limitations of natural selection,[1] the theologian that his treatment of the problems of sin and suffering was inadequate or at least unorthodox, the social scientist that he failed to take sufficient account of the facts of political and social history. But he saw that what was needed at the moment was a broad sweep and a comprehensive treatment. This was what he essayed in *The Phenomenon of Man*. In my view he achieved a remarkable success, and opened up vast territories of thought to further exploration and detailed mapping.

The facts of Père Teilhard's life help to illuminate the development of his thought. His father was a small landowner in Auvergne, a gentleman farmer who was also an archivist, with a taste for natural history. Pierre was born in 1881, the fourth in a family of eleven. At the age of ten he went as a boarder to a Jesuit College where, besides doing well in all prescribed subjects of study, he became devoted to field geology and mineralogy. When eighteen years old, he decided to become a Jesuit, and entered their order. At the age of twenty-four, after

[1] Though in his Institute for Human Studies he envisaged a section of Eugenics.

an interlude in Jersey mainly studying philosophy, he was sent to teach physics and chemistry in a Jesuit College at Cairo. In the course of his three years in Egypt, and a further four studying theology in Sussex, he acquired real competence in geology and palaeontology ; and before being ordained priest in 1912, a reading of Bergson's *Evolution Créatrice* had helped to inspire in him a profound interest in the general facts and theories of evolution. Returning to Paris, he pursued his geological studies and started working under Marcellin Boule, the leading prehistorian and archaeologist of France, in his Institute of Human Palaeontology at the Museum of Natural History. It was here that he met his lifelong friend and colleague in the study of prehistory, the Abbé Breuil, and that his interests were first directed to the subject on which his life's work was centred—the evolution of man. In 1913 he visited the site where the famous (and now notorious) Piltdown skull had recently been unearthed, in company with its discoverer Dr. Dawson and the leading English palaeontologist Sir Arthur Smith Woodward. This was his first introduction to the excitements of palaeontological discovery and scientific controversy.

During the first World War he served as a stretcher-bearer, receiving the Military Medal and the Legion of Honour, and learnt a great deal about his fellow men and about his own nature. The war strengthened his sense of religious vocation, and in 1918 he made a triple vow of poverty, chastity and obedience.

By 1919 the major goals of his life were clearly indicated. Professionally, he had decided to embark on a geological career, with special emphasis on palaeontology. As a thinker, he had reached a point where the entire phenomenal universe, including man, was revealed as a process of evolution, and he found himself impelled to build up a generalised theory or philosophy of evolutionary process which would take account of human history and human personality as well as of biology, and from which one could draw conclusions as to the future evolution of man on earth. And as a dedicated Christian priest, he felt it imperative to try to reconcile Christian theology with this evolutionary philosophy, to relate the facts of religious experience to those of natural science.

Returning to the Sorbonne, he took his Doctorate in 1922. He had already become Professor of Geology at the Catholic Institute of Paris, where his lectures attracted great attention among the students (three of whom are now teaching in the University of Paris). In 1923, however, he went to China for a year on behalf of the Museum, on a palaeontological mission directed by another Jesuit, Père Licent. His *Lettres de Voyage* reveal the impression made on him by the voyage through the tropics, and by his first experience of geological research in the desert remoteness of Mongolia and north-western China. This expedition inspired *La Messe sur le Monde*, a remarkable and truly poetical essay which was at one and the same time mystical and realistic, religious and philosophical.

A shock awaited him after his return to France. Some of the ideas which he had expressed in his lectures about original sin and its relation to evolution, were regarded as unorthodox by his religious superiors, and he was forbidden to continue teaching. In 1926 he returned to work with Père Licent in China, where he was destined to stay, with brief returns to France and excursions to the United States, to Abyssinia, India, Burma and Java, for twenty years. Here, as scientific adviser to the Geological Survey of China, centred first at Tientsin and later at Peking, he met and worked with outstanding palaeontologists of many nations, and took part in a number of expeditions, including the Citroën *Croisière Jaune* under Haardt, and Davidson Black's expedition which unearthed the skull of Peking man.

In 1938 he was appointed Director of the Laboratory of Advanced Studies in Geology and Palaeontology in Paris, but the outbreak of war prevented his return to France. His enforced isolation in China during the six war years, painful and depressing though it often was, undoubtedly helped his inner spiritual development (as the isolation of imprisonment helped to mature the thought and character of Nehru and many other Indians).

It encouraged ample reading and reflection, and stimulated the full elaboration of his thought.

It was a nice stroke of irony that the action of Père Teilhard's religious superiors in barring him from teaching in France because of his ideas on human evolution, should have led him to China and brought him into intimate association with one of the most important discoveries in that field, and driven him to enlarge and consolidate his 'dangerous thoughts'.

During the whole of this period he was writing essays and books on various aspects and implications of evolution, culminating in 1938 in the manuscript of *Le Phénomène Humain*. But he never succeeded in obtaining permission to publish any of his controversial or major works. This caused him much distress, for he was conscious of a prophetic mission : but he faithfully observed his vow of obedience. Professionally too he was extremely active throughout this period. He contributed a great deal to our knowledge of palaeolithic cultures in China and neighbouring areas, and to the general understanding of the geology of the Far East. This preoccupation with large-scale geology led him to take an interest in the geological development of the world's continents : each continent, he considered, had made its own special contribution to biological evolution. He also did important palaeontological work on the evolution of various mammalian groups.

The wide range of his vision made him impatient of over-specialisation, and of the timidity which refuses to pass from detailed study to broad synthesis. With his conception of mankind as at the same time an unfinished product of past evolution and an agency of distinctive evolution to come, he was particularly impatient of what he felt as the narrowness of those anthropologists who limited themselves to a study of physical structure and the details of primitive social life. He wanted to deal with the entire human phenomenon, as a transcendence of biological by psychosocial evolution. And he had considerable success in redirecting along these lines the institutions with which he was connected.

Back in France in 1946, Père Teilhard plunged eagerly into European intellectual life, but in 1947 he had a serious heart attack, and was compelled to spend several months convalescing in the country. On his return to Paris, he was enjoined by his superiors not to write any more on philosophical subjects : and in 1948 he was forbidden to put forward his candidature for a Professorship in the Collège de France in succession to the Abbé Breuil, though it was known that this, the highest academic position to which he could aspire, was open to him. But perhaps the heaviest blow awaited him in 1950, when his application for permission to publish *Le Groupe Zoologique Humain* (a recasting of *Le Phénomène Humain*) was refused in Rome. By way of compensation he was awarded the signal honour of being elected *Membre de l'Institut*, as well as having previously become a Corresponding Member of the *Académie des Sciences*, an officer of the *Légion d'Honneur*, and a director of research in the *Centre National de la Recherche Scientifique*.

Already in 1948 he had been invited to visit the U.S.A., where he made his first contacts with the Wenner-Gren Foundation (or Viking Foundation as it was then called), in whose friendly shelter he spent the last four years of his life. The Wenner-Gren Foundation also sponsored his two visits to South Africa, where he was able to study at first hand the remarkable discoveries of Broom and Dart concerning *Australopithecus*, that near-ancestor of man, and to lay down a plan for the future co-ordination of palaeontological and archaeological work in this area, so important as a centre of hominid evolution.

His position in France became increasingly difficult, and in 1951 he moved his headquarters to New York. Here, at the Wenner-Gren Foundation, he played an important role in framing anthropological policy, and made valuable contributions to the international symposia which it organised. And here, in 1954, I had the privilege of working with him in one of the remarkable discussion groups set up as part of the Columbia Bicentennial celebrations. Just before this, he had returned to France for a brief but stimulating month of discussion.

Throughout this period, he had been actively developing his ideas, and had written his spiritual autobiography, *Le Cœur de la Matière*, the semi-technical *Le Groupe Zoologique Humain*, and various technical and general articles later included in the collections entitled *La Vision du Passé* and *L'Apparition de l'Homme*.

He was prevailed on to leave his manuscripts to a friend. They therefore could be published after his death, since permission to publish is only required for the work of a living writer. The prospect of eventual publication must have been a great solace to him, for he certainly regarded his general and philosophical writings as the keystone of his life's work, and felt it his supreme duty to proclaim the fruits of his labour.

It was my privilege to have been a friend and correspondent of Père Teilhard for nearly ten years ; and it is my privilege now to introduce this, his most notable work, to English-speaking readers.

His influence on the world's thinking is bound to be important. Through his combination of wide scientific knowledge with deep religious feeling and a rigorous sense of values, he has forced theologians to view their ideas in the new perspective of evolution, and scientists to see the spiritual implications of their knowledge. He has both clarified and unified our vision of reality. In the light of that new comprehension, it is no longer possible to maintain that science and religion must operate in thought-tight compartments or concern separate sectors of life ; they are both relevant to the whole of human existence. The religiously-minded can no longer turn their backs upon the natural world, or seek escape from its imperfections in a supernatural world ; nor can the materialistically-minded deny importance to spiritual experience and religious feeling.

Like him, we must face the phenomena. If we face them resolutely, and avail ourselves of the help which his intellectual and spiritual travail has provided, we shall find a more assured basis for our thought and a more certain direction for our evolutionary advance. But, like him, we must not take refuge in abstractions of generalities. He always took account of the specific realities of man's present situation, though set against the more general realities of long-term evolution ; and he always endeavoured to think concretely, in terms of actual patterns of organisation—their development, their mode of operation and their effects.

As a result, he has helped us to define more adequately both our own nature, the general evolutionary process, and our place and role in it. Thus clarified, the evolution of life becomes a comprehensible phenomenon. It is an anti-entropic process, running counter to the second law of thermodynamics with its degradation of energy and its tendency to uniformity. With the aid of the sun's energy, biological evolution marches uphill, producing increased variety and higher degrees of organisation.

It also produces more varied, more intense and more highly organised mental activity or awareness. During evolution, awareness (or if you prefer, the mental properties of living matter) becomes increasingly important to organisms, until in mankind it becomes the most important characteristic of life, and gives the human type its dominant position.

After this critical point has been passed, evolution takes on a new character : it becomes primarily a psychosocial process, based on the cumulative transmission of experience and its results, and working through an organised system of awareness, a combined operation of knowing, feeling and willing. In man, at least during the historical and proto-historical periods, evolution has been characterised more by cultural than by genetic or biological change.

On this new psychosocial level, the evolutionary process leads to new types and higher degrees of organisation. On the one hand there are new patterns of co-operation among individuals—co-operation for practical control, for enjoyment, for education, and notably in the last few centuries, for obtaining new knowledge ; and on the other there are new patterns of thought, new organisations of awareness and its products.

As a result, new and often wholly unexpected possibilities have been realised, the variety and degree of human fulfilment has been increased. Père Teilhard enables us to see which possibilities are in the long run desirable. What is more, he has helped to define the conditions of advance, the conditions which will permit an increase of fulfilment and prevent an increase of frustration. The conditions of advance are these : global unity of mankind's noetic organisation or system of awareness, but a high degree of variety within that unity ; love, with goodwill and full co-operation ; personal integration and internal harmony ; and increasing knowledge.

Knowledge is basic. It is knowledge which enables us to understand the world and ourselves, and to exercise some control or guidance. It sets us in a fruitful and significant relation with the enduring processes of the universe. And, by revealing the possibilities of fulfilment that are still open, it provides an overriding incentive.

We, mankind, contain the possibilities of the earth's immense future, and can realise more and more of them on condition that we increase our knowledge and our love. That, it seems to me, is the distillation of *The Phenomenon of Man*.

*London, December 1958*

# *Preface*

IF THIS book is to be properly understood, it must be read not as a work on metaphysics, still less as a sort of theological essay, but purely and simply as a scientific treatise. The title itself indicates that. This book deals with man *solely* as a phenomenon ; but it also deals with the *whole* phenomenon of man.

In the first place, it deals with man *solely* as a phenomenon. The pages which follow do not attempt to give an explanation of the world, but only an introduction to such an explanation. Put quite simply, what I have tried to do is this ; I have chosen man as the centre, and around him I have tried to establish a coherent order between antecedents and consequents. I have not tried to discover a system of ontological and causal relations between the elements of the universe, but only an experimental law of recurrence which would express their successive appearance in time. Beyond these first purely *scientific* reflections, there is obviously ample room for farther-reaching speculations of the philosopher and the theologian. Of set purpose, I have at all times carefully avoided venturing into that field of the essence of being. At most I am confident that, on the plane of experience, I have identified with some accuracy the combined movement towards unity, and have marked the places where philosophical and religious thinkers, in pursuing the matter further, would be entitled, for reasons of a higher order, to look for breaches of continuity.[1]

But this book also deals with the *whole* phenomenon of man. Without contradicting what I have just said (however much it may appear to do so) it is this aspect which might possibly make my suggestions *look* like a philosophy. During the last fifty years

[1] See, for example, the footnotes on pp. 169, 186, 298.

or so, the investigations of science have proved beyond all doubt that there is no fact which exists in pure isolation, but that every experience, however objective it may seem, inevitably becomes enveloped in a complex of assumptions as soon as the scientist attempts to express it in a formula. But while this aura of subjective interpretation may remain imperceptible where the field of observation is limited, it is bound to become practically dominant as soon as the field of vision extends to the whole. Like the meridians as they approach the poles, science, philosophy and religion are bound to converge as they draw nearer to the whole. I say 'converge' advisedly, but without merging, and without ceasing, to the very end, to assail the real from different angles and on different planes. Take any book about the universe written by one of the great modern scientists, such as Poincaré, Einstein or Jeans, and you will see that it is impossible to attempt a general scientific interpretation of the universe without *giving the impression* of trying to explain it through and through. But look a little more closely and you will see that this 'hyperphysics' is still not a metaphysic.

In the course of every effort of this kind to give a scientific description of the whole, it is natural that certain basic assumptions, on which the whole further structure rests, should make their influence felt to the fullest possible extent. In the specific instance of the present Essay, I think it important to point out that two basic assumptions go hand in hand to support and govern every development of the theme. The first is the primacy accorded to the psychic and to thought in the stuff of the universe, and the second is the 'biological' value attributed to the social fact around us.

The pre-eminent significance of man in nature, and the organic nature of mankind; these are two assumptions that one may start by trying to reject, but without accepting them, I do not see how it is possible to give a full and coherent account of the phenomenon of man.

*Paris, March 1947*

# *Foreword*

## SEEING

THIS WORK may be summed up as an attempt *to see* and *to make others see* what happens to man, and what conclusions are forced upon us, when he is placed fairly and squarely within the framework of phenomenon and appearance.

Why should we want to see, and why in particular should we single out man as our object?

*Seeing*. We might say that the whole of life lies in that verb—if not ultimately, at least essentially. Fuller being is closer union: such is the kernel and conclusion of this book. But let us emphasise the point: union increases only through an increase in consciousness, that is to say in vision. And that, doubtless, is why the history of the living world can be summarised as the elaboration of ever more perfect eyes within a cosmos in which there is always something more to be seen. After all, do we not judge the perfection of an animal, or the supremacy of a thinking being, by the penetration and synthetic power of their gaze? To try to see more and better is not a matter of whim or curiosity or self-indulgence. *To see or to perish* is the very condition laid upon everything that makes up the universe, by reason of the mysterious gift of existence. And this, in superior measure, is man's condition.

But if it is true that it is so vital and so blessed to *know*, let us ask again why we are turning our attention particularly to man. Has man not been adequately described already, and is he not a tedious subject? Is it not precisely one of the attractions of science that it rests our eyes by turning them away from man?

Man has a double title, as the twofold centre of the world, to impose himself on our effort to see, as the key to the universise.

Subjectively, first of all, we are inevitably the centre of perspective of our own observation. In its early, naïve stage, science, perhaps inevitably, imagined that we could observe phenomena in themselves, as they would take place in our absence. Instinctively physicists and naturalists went to work as though they could look down from a great height upon a world which their consciousness could penetrate without being submitted to it or changing it. They are now beginning to realise that even the most objective of their observations are steeped in the conventions they adopted at the outset and by forms or habits of thought developed in the course of the growth of research ; so that, when they reach the end of their analyses they cannot tell with any certainty whether the structure they have reached is the essence of the matter they are studying, or the reflection of their own thought. And at the same time they realise that as the result of their discoveries, they are caught body and soul to the network of relationships they thought to cast upon things from outside : in fact they are caught in their own net. A geologist would use the words metamorphism and endomorphism. Object and subject marry and mutually transform each other in the act of knowledge ; and from now on man willy-nilly finds his own image stamped on all he looks at.

This is indeed a form of bondage, for which, however, a unique and assured grandeur provides immediate compensation.

It is tiresome and even humbling for the observer to be thus fettered, to be obliged to carry with him everywhere the centre of the landscape he is crossing. But what happens when chance directs his steps to a point of vantage (a cross-roads, or intersecting valleys) from which, not only his vision, but things themselves radiate? In that event the subjective viewpoint coincides with the way things are distributed objectively, and perception reaches its apogee. The landscape lights up and yields its secrets. He sees.

That seems to be the privilege of man's knowledge.

It is not necessary to be a man to perceive surrounding things and forces 'in the round'. All the animals have reached this point as well as us. But it is peculiar to man to occupy a position in nature at which the convergent lines are not only visual but structural. The following pages will do no more than verify and analyse this phenomenon. By virtue of the quality and the biological properties of thought, we find ourselves situated at a singular point, at a ganglion which commands the whole fraction of the cosmos that is at present within reach of our experience. Man, the centre of perspective, is at the same time the *centre of construction* of the universe. And by expediency no less than by necessity, all science must be referred back to him. If to see is really to become more, if vision is really fuller being, then we should look closely at man in order to increase our capacity to live.

But to do this we must focus our eyes correctly.

From the dawn of his existence, man has been held up as a spectacle to himself. Indeed for tens of centuries he has looked at nothing but himself. Yet he has only just begun to take a scientific view of his own significance in the physical world. There is no need to be surprised at this slow awakening. It often happens that what stares us in the face is the most difficult to perceive. The child has to learn to separate out the images which assail the newly-opened retina. For man to discover man and take his measure, a whole series of 'senses' have been necessary, whose gradual acquisition, as we shall show, covers and punctuates the whole history of the struggles of the mind :

A sense of spatial immensity, in greatness and smallness, disarticulating and spacing out, within a sphere of indefinite radius, the orbits of the objects which press round us ;

A sense of depth, pushing back laboriously through endless series and measureless distances of time, which a sort of sluggishness of mind tends continually to condense for us in a thin layer of the past ;

A sense of number, discovering and grasping unflinchingly the bewildering multitude of material or living elements involved in the slightest change in the universe ;

A sense of proportion, realising as best we can the difference of physical scale which separates, both in rhythm and dimension, the atom from the nebula, the infinitesimal from the immense ;

A sense of quality, or of novelty, enabling us to distinguish in nature certain absolute stages of perfection and growth, without upsetting the physical unity of the world ;

A sense of movement, capable of perceiving the irresistible developments hidden in extreme slowness—extreme agitation concealed beneath a veil of immobility—the entirely new insinuating itself into the heart of the monotonous repetition of the same things ;

A sense, lastly, of the organic, discovering physical links and structural unity under the superficial juxtaposition of successions and collectivities.

Without these qualities to illuminate our vision, man will remain indefinitely for us—whatever is done to make us see—what he still represents to so many minds : an erratic object in a disjointed world. Conversely, we have only to rid our vision of the threefold illusion of smallness, plurality and immobility, for man effortlessly to take the central position we prophesied—the momentary summit of an anthropogenesis which is itself the crown of a cosmogenesis.

Man is unable to see himself entirely unrelated to mankind, neither is he able to see mankind unrelated to life, nor life unrelated to the universe.

Thence stems the basic plan of this work : *Pre-Life : Life : Thought*—three events sketching in the past and determining for the future (*Survival*) a single and continuing trajectory, the curve of the phenomenon of man.

The phenomenon of man—I stress this.

This phrase is not chosen at random, but for three reasons.

First to assert that man, in nature, is a genuine fact falling (at least partially) within the scope of the requirements and methods of science ;

Secondly, to make plain that of all the facts offered to our knowledge, none is more extraordinary or more illuminating ;

Thirdly, to stress the special character of the Essay I am presenting.

I repeat that my only aim, and my only vantage-ground in these pages, is to try to see ; that is to say, to try to develop a *homogeneous* and *coherent* perspective of our general extended experience of man. A *whole* which unfolds.

So please do not expect a final explanation of things here, nor a metaphysical system. Neither do I want any misunderstanding about the degree of reality which I accord to the different parts of the film I am projecting. When I try to picture the world before the dawn of life, or life in the Palaeozoic era, I do not forget that there would be a cosmic contradiction in imagining a man as spectator of those phases which ran their course before the appearance of thought on earth. I do not pretend to describe them as they really were, but rather as we must picture them to ourselves so that the world may be true for us at this moment. What I depict is not the past *in itself*, but as it must appear to an observer standing on the advanced peak where evolution has placed us. It is a safe and modest method and yet, as we shall see, it suffices, through symmetry, to bring out ahead of us surprising visions of the future.

Even reduced to these humble proportions, the views I am attempting to put forward here are, of course, largely tentative and personal. Yet inasmuch as they are based on arduous investigation and sustained reflection, they give an idea, by means of one example, of the way in which the problem of man presents itself in science today.

When studied narrowly in himself by anthropologists or jurists, man is a tiny, even a shrinking, creature. His over-pronounced individuality conceals from our eyes the whole to which he belongs ; as we look at him our minds incline to break nature up into pieces and to forget both its deep inter-relations and its measureless horizons : we incline to all that is bad in anthropocentrism. And it is this that still leads scientists to refuse to consider man as an object of scientific scrutiny except through his body.

The time has come to realise that an interpretation of the universe—even a positivist one—remains unsatisfying unless it covers the interior as well as the exterior of things ; mind as well as matter. The true physics is that which will, one day, achieve the inclusion of man in his wholeness in a coherent picture of the world.

I hope I shall persuade the reader that such an attempt is possible, and that the preservation of courage and the joy of action in those of us who wish, and know how, to plumb the depths of things, depend on it.

In fact I doubt whether there is a more decisive moment for a thinking being than when the scales fall from his eyes and he discovers that he is not an isolated unit lost in the cosmic solitudes, and realises that a universal will to live converges and is hominised in him.

In such a vision man is seen not as a static centre of the world —as he for long believed himself to be—but as the axis and leading shoot of evolution, which is something much finer.

BOOK ONE

# BEFORE LIFE CAME

CHAPTER ONE

# THE STUFF OF THE UNIVERSE

To PUSH anything back into the past is equivalent to reducing it to its simplest elements. Traced as far as possible in the direction of their origins, the last fibres of the human aggregate are lost to view and are merged in our eyes with the very stuff of the universe.

As for the stuff of the universe—the ultimate residue of the ever more advanced analyses of science—I have not cultivated that direct and familiar contact with it which would enable me to do it justice, that contact which comes from experiment and not from reading and makes all the difference. Besides, I know the danger of trying to construct a lasting edifice with hypotheses which are only expected to last for a day, even in the minds of those who originate them.

To a considerable extent, the representation of the atom accepted at this moment is nothing more than a simple means, graphic even while subject to revision, enabling the scientist to put together and to show the non-contradiction of the ever more various 'effects' manifested by matter—many of which, moreover, have still no recognisable prolongation in man.

As I am a naturalist rather than a physicist, obviously I shall avoid dealing at length with or placing undue reliance upon these complicated and fragile edifices.

On the other hand, among the variety of overlapping theories, a certain number of characteristics emerge which are inevitable in any suggested explanation of the universe. It is of these 'imposed' factors that it is not unbecoming for a naturalist to speak when engaged on a general study of the phenomenon of man. In fact, inasmuch as they express the conditions belonging to all natural change, even biological, he is bound to take them as his point of departure.

## 1. ELEMENTAL MATTER

Observed from this special angle, and considered at the outset in its elemental state (by which I mean at any moment, at any point, and in any volume), the stuff of tangible things reveals itself with increasing insistence as radically particulate yet essentially related, and lastly, prodigiously active.

Plurality, unity, energy : the three faces of matter.

### A. *Plurality*

The profoundly ' atomic '[1] character of the universe is visible in everyday experience, in raindrops and grains of sand, in the hosts of the living, and the multitude of stars ; even in the ashes of the dead. Man has needed neither microscope nor electronic analysis in order to suspect that he lives surrounded by and resting on dust. But to count the grains and describe them, all the patient craft of modern science was necessary. The atoms of Epicurus were inert and indivisible. And the infinitesimal worlds of Pascal could still possess their animalcules. Today we have gone far beyond such instinctive or inspired guesswork both in certainty and precision. The scaling down is unlimited. Like the tiny diatom shells whose markings, however magnified, change almost indefinitely into new patterns, so each particle of matter, ever smaller and smaller, under the physicist's analysis tends to reduce itself into something yet more finely granulated. And at each new step in this progressive approach to the infinitely small the whole configuration of the world is for a moment blurred and then renewed.

[1] [*Atomicité.*]

When we probe beyond a certain degree of depth and dilution, the familiar properties of our bodies—light, colour, warmth, impenetrability, etc.—lose their meaning.

Indeed our sensory experience turns out to be a floating condensation on a swarm of the undefinable. Bewildering in its multiplicity and its minuteness, the substratum of the tangible universe is in an unending state of disintegration as it goes downward.

### B. *Unity*

On the other hand the more we split and pulverise matter artificially, the more insistently it proclaims its *fundamental unity*.

In its most imperfect form, but the simplest to imagine, this unity reveals itself in the astonishing similarity of the elements met with. Molecules, atoms, electrons—whatever the name, whatever the scale—these minute units (at any rate when viewed from our distance) manifest a perfect identity of mass and of behaviour. In their dimensions and actions they seem astonishingly calibrated—and monotonous. It is almost as if all that surface play which charms our lives tends to disappear at deeper levels. It is almost as if the stuff of which all stuff is made were reducible in the end to some simple and unique kind of substance.

Thus the *unity of homogeneity*. To the cosmic corpuscles we should find it natural to attribute an individual radius of action as limited as their dimensions. We find, on the contrary, that each of them can only be defined by virtue of its influence on all around it. Whatever space we suppose it to be in, each cosmic element radiates in it and entirely fills it. However narrowly the 'heart' of an atom may be circumscribed, its realm is co-extensive, at least potentially, with that of every other atom. This strange property we will come across again, even in the human molecule.

We add : *collective unity*. The innumerable foci which share a given volume of matter are not therefore independent of each other. Something holds them together. Far from behaving as a mere inert receptacle, the space filled by their multitude operates upon it like an active centre of direction and transmission in which their plurality is organised. We do not get what we call matter as a result of the simple aggregation and juxtaposition of atoms. For that, a mysterious identity must absorb and cement them, an influence at which our mind rebels in bewilderment at first but which in the end it must perforce accept.

We mean the sphere above the centres and enveloping them.

Throughout these pages, in each new phase of anthropogenesis, we shall find ourselves faced by the unimaginable reality of collective bonds, and we shall have to struggle with them without ceasing until we succeed in recognising and defining their true nature. Here in the beginning it is sufficient to include them all under the empirical name given by science to their common initial principle, namely *energy*.

### C. *Energy*

Under this name, which conveys the experience of effort with which we are familiar in ourselves, physics has introduced the precise formulation of a capacity for action or, more exactly, for interaction. Energy is the measure of that which passes from one atom to another in the course of their transformations. A unifying power, then, but also, because the atom appears to become enriched or exhausted in the course of the exchange, the expression of structure.

From the aspect of energy, renewed by radio-active phenomena, material corpuscles may now be treated as transient reservoirs of concentrated power. Though never found in a state of purity, but always more or less granulated (even in light) energy nowadays represents for science the most primitive form of universal stuff. Hence we find our minds instinctively tending to represent energy as a kind of homogeneous, primordial flux in which all that has shape in the world is but a series of fleeting 'vortices'. From this point of view, the universe would find its stability and final unity *at the end of its decomposition. It would be held together from below.*

Let us keep the discoveries and indisputable measurements of physics. But let us not become bound and fettered to the perspective of final equilibrium that they seem to suggest. A more complete study of the movements of the world will oblige us, little by little, to turn it upside down ; in other words, to discover that if things hold and hold together, it is only by reason of complexity, *from above*.

## 2. TOTAL MATTER

Up to now we have been looking at matter as such, that is to say according to its qualities and in any given volume—as though it were permissible for us to break off a fragment and study this sample apart from the rest. It is time to point out that this procedure is merely an intellectual dodge. Considered in its physical, concrete reality, the stuff of the universe cannot divide itself but, as a kind of gigantic 'atom', it forms in its totality (apart from thought on which it is centred and concentrated at the other end) the only real indivisible. The history of consciousness and its place in the world remain incomprehensible to anyone who has not seen first of all that the cosmos in which man finds himself caught up constitutes, by reason of the unimpeachable wholeness of its whole, a *system*, a *totum* and a *quantum* : a system by its plurality, a totum by its unity, a quantum by its energy ; all three within a boundless contour.

Let us try to make this clear.

### A. *The System*

The existence of 'system' in the world is at once obvious to every observer of nature, no matter whom.

The arrangement of the parts of the universe has always been a source of amazement to men. But this disposition proves itself more and more astonishing as, every day, our science is able to make a more precise and penetrating study of the facts. The farther and more deeply we penetrate into matter, by means of increasingly powerful methods, the more we are confounded by the interdependence of its parts. Each element of the cosmos is positively woven from all the others : from beneath itself by the mysterious phenomenon of 'composition', which makes it subsistent through the apex of an organised whole; and from above through the influence of unities of a higher order which incorporate and dominate it for their own ends.

It is impossible to cut into this network, to isolate a portion without it becoming frayed and unravelled at all its edges.

All around us, as far as the eye can see, the universe holds together, and only one way of considering it is really possible, that is, to take it as a whole, in one piece.

### B. *The Totum*

Now, if we consider this whole more attentively, we quickly see that it is something quite other than a mere entanglement of articulated inter-connections. If one says fabric or network, one thinks of a homogeneous plexus of similar units which it may indeed be impossible to section, but of which it is sufficient to have recognised the basic unit and to have defined the law to be able to understand the whole by repetition : a crystal or arabesque whose laws are valid for whatever space it fills, but which is wholly contained in a single mesh.

Between such a structure and the structure of matter there is nothing in common.

In its different orders of magnitude, matter never repeats its different combinations. For expedience and simplicity we sometimes like to imagine the world as being a series of planetary systems superimposed, the one on the other, and grading from the infinitely small to the infinitely big : Pascal's two abysses once again. This is only an illusion. The envelopes composing matter are thoroughly heterogeneous the one with regard to the other. First we have a vague circle of electrons and other inferior units ; then a better-defined circle of simple bodies in which the elements are distributed as periodic functions of the atom of hydrogen ; farther on another circle, of inexhaustible molecular combinations ; and lastly, jumping or recoiling from the infinitesimal to the infinite, a circle of stars and galaxies. These multiple zones of the cosmos envelop without imitating each other in such a way that we cannot pass from one to another by a simple change of coefficients. Here is no repetition of the same theme on a different scale. The order and the design do not appear except in the whole. The mesh of the universe is the universe itself.

Thus it is not enough merely to assert that matter forms a block or whole.

The stuff of the universe, woven in a single piece according to one and the same system,[1] but never repeating itself from one point to another, represents a single figure. Structurally, it forms a Whole.

### C. *The Quantum*

Now, if the natural unity of concrete space indeed coincides with the totality of space itself, we must try to re-define energy with reference to space as a whole.

This leads us to two conclusions.

The first is that the radius of action proper to each cosmic element must be prolonged in theory to the utmost limits of the world itself. As we said above, since the atom is naturally co-extensive with the whole of the space in which it is situated—and since, on the other hand, we have just seen that a universal space is *the only space there is*—we are bound to admit that this immensity represents the sphere of action common to all atoms. The volume of each of them is the volume of the universe. The

[1] Which we shall call later on 'the Law of Consciousness and Complexity'.

atom is no longer the microscopic, closed world we may have imagined to ourselves. It is the infinitesimal centre of the world itself.

Now, on the other hand, let us turn our attention to the entirety of the infinitesimal centres which share the universal sphere among themselves. Indefinite though their number may be, they constitute in their multitude a group which has precise effects. For the whole, because it exists, must express itself in a global capacity for action of which we find the partial resultant in each one of us. Thus we find ourselves led on to envisage and conceive a dynamic standard of the world.

True the world has apparently limitless contours. To use varying metaphors: it behaves to our senses, either as a progressively attenuated environment which vanishes without a limital surface in an infinitely decreasing gradation, or as a curved and closed space within which all the lines of our experience turn back upon themselves, in which case matter only appears boundless to us because we cannot emerge from it.

This is no reason for refusing it a quantum of energy, which the physicists, incidentally, already think they are in a position to measure.

But this quantum only takes on its full significance when we try to define it with regard to a concrete natural movement—that is to say, in *duration*.

## 3. THE EVOLUTION OF MATTER

Physics was born, in the last century, under the double sign of fixity and geometry. Its ideal, in its youth, was to find a mathematical explanation of a world imagined as a system of stable elements in a closed equilibrium. Then, following all science of the real, it was inevitably drawn by its own progress into becoming a history. Today, positive knowledge of things is identified with the study of their development. Farther on, in the chapter on Thought, we shall have to describe and interpret the vital revolution in human consciousness brought about by the quite modern discovery of duration. Here we need only ask ourselves how our views about matter are enlarged by the introduction of this new dimension.

In essence, the change wrought in our experience by the appearance of what we shall soon call space-time is this, that everything that up to then we regarded and treated as points in our cosmological constructions became instantaneous sections of indefinite temporal fibres. To our opened eyes each element of things is henceforth extended backwards (and tends to be continued forwards) as far as the eye can see in such a way that the entire spatial immensity is no more than a section 'at the time *t*' of a trunk whose roots plunge down into the abyss of an unfathomable past, and whose branches rise up somewhere to a future that, at first sight, has no limit. In this new perspective the world appears like a mass in process of transformation. The universal totum and quantum tend to express and define themselves in cosmogenesis. What at this moment are the appearance (qualitative) assumed from the point of view of the physicists and the rules followed (quantitative) by this evolution of matter ?

## A. *The Appearance*

As seen in its central portion, which is the most distinct, the evolution of matter, in current theory, comes back to the gradual building up by growing complication of the various elements recognised by physical chemistry. To begin with, at the very bottom there is a still unresolved simplicity, luminous in nature and not to be defined in terms of figures. Then, suddenly( ?)[1]

[1] Some years ago this first birth of the corpuscles was imagined rather as a sudden *condensation* (as in a saturated environment) of a primordial substance or stuff, diffused throughout limitless space. Nowadays, for various convergent reasons, notably Relativity combined with the centrifugal retreat of the galaxies, physicists prefer to turn to the idea of an *explosion* pulverising a primitive quasi-atom within which space-time would be strangulated (in a

came a swarming of elementary corpuscles, both positive and negative (protons, neutrons, electrons, photons) : the list increases incessantly. Then the harmonic series of simple bodies, strung out from hydrogen to uranium on the notes of the atomic scale. Next follows the immense variety of compound bodies in which the molecular weights go on increasing up to a certain critical value above which, as we shall see, we pass on to life. There is not one term in this long series but must be regarded, from sound experimental proofs, as being composed of nuclei and electrons. This fundamental discovery that all bodies owe their origin to arrangements of a single initial corpuscular type is the beacon that lights the history of the universe to our eyes. In its own way, matter has obeyed from the beginning that great law of biology to which we shall have to recur time and time again, the law of ' complexification '.[1]

I say in its own way because, at the stage of the atom, we are still ignorant of many points in the history of the world.

First of all, must all the elements mount each successive rung of the ladder from the most simple to the most complicated by a kind of onto- or phylo-genesis in order to raise themselves in the series of simple bodies ? Or do the atomic numbers only represent a rhythmic series of states of equilibrium, sets of pigeon-holes, as it were, into which nuclei and electrons fall in rough assemblages ? Moreover, in the one instance as in the other, must we regard the various combinations of nuclei as being equally possible at any one time ? Or, on the other hand, must we suppose that on the whole, statistically, the heavy atoms only appear in a determinate order, after the lighter ones ?

[1] [*Complexification* in the original: taken over here as the substantival form of the very rare English verb ' complexify '—to make complex.]

---

sort of natural absolute zero) at only some milliards of years behind us. For understanding the following pages, the two hypotheses are equivalent, in the sense that they put us, the one just as much as the other, in the midst of a corpuscular multitude from which we cannot escape in any direction; neither round about nor behind—but possibly forwards (cf. Part 4, chapter 2) through a singular point of interiorisation.

It does not appear that science is at present able to give definitive answers to these questions, or to others like them. At the present time we are less well informed about the ascending evolution of atoms (I do not say ‘ the disintegration ’) than we are about the pre-living and living molecules. It is none the less true, and this is the only point of real importance that concerns us here, that from its most distant formulations matter reveals itself to us *in a state of genesis* or becoming—this genesis allowing us to distinguish two of the aspects most characteristic of it in its subsequent stages. First of all, to begin with a critical phase, that of *granulation*, which abruptly and once and for all gave birth to the constituents of the atom and perhaps to the atom itself. Next, at least from the molecular level, of going on additively by a process of growing complexity.

Everything does not happen continuously at any one moment in the universe. Neither does everything happen everywhere in it.

So we may summarise in a few lines the ideas about the transformations of matter accepted by science today : but only by considering the latter in their temporal succession, and without as yet putting them anywhere within the cosmic expanse. Historically, the stuff of the universe goes on becoming concentrated into ever more organised forms of matter. But *where*, then, do these metamorphoses take place, beginning, let us say, with the framework of molecules ? Is it indifferently at any point in space ? Not at all, as we all know, but only in the heart and on the surface of the stars. From having considered the infinitely small elements we are abruptly compelled to raise our eyes to infinitely great sidereal masses.

The sidereal masses . . . Our science is at the same time troubled and fascinated by these colossal unities, which in some ways behave like atoms, but whose constitution baffles us by its enormous and—in appearance only ?—irregular complexity. Perhaps the day will come when some arrangement or periodicity will become apparent in the stellar distribution both as regards their composition and their position. Do not a ‘ stratigraphy ’ and a 'chemistry' of the heavens inevitably extend the story of the atoms ?

We have not to entangle ourselves in these still misty perspectives. No matter how fascinating they may be, they surround man rather than lead up to him. On the other hand, because of its consequences even up to the genesis of the intellect, we must notice and record the definite connection which, genetically, associates the atom with the star. For a long time yet physics may hesitate over the structure to be assigned to the astral immensities. In the meantime one thing is certain and is enough to guide our steps along the ways of anthropogenesis. That is that the making of greater material complexes can only take place under cover of a previous concentration of the stuff of the universe in nebulae and suns. Whatever the overall figure of the worlds may be, the chemical function of each one of them already has a definable meaning for us. The stars are laboratories in which the evolution of matter proceeds in the direction of large molecules, and that according to determinate quantitative rules which we must now discuss.

### B. *The Numerical Laws*

What ancient thought half perceived and imagined as a natural harmony of numbers, modern science has grasped and realised in the precision of formulae dependent on measurement. Indeed, we owe our knowledge of the macro-structure and micro-structure of the universe far more to increasingly accurate measurements than to direct observations. And, again, it is ever bolder measurements that have revealed to us the calculable conditions to which every transformation of matter is subject according to the force it calls into play.

This is not the place for me to embark on a critical discussion of the laws of energy. That part of them that is indispensable and accessible to every world-historian may be simply summarised. Considered from this biological aspect, broadly speaking, they may be reduced to the two following principles :

*First Principle.* During changes of a physico-chemical type we do not detect any measurable emergence of new energy.

Every synthesis costs something. That is a fundamental condition of things which persists, as we know, even into the spiritual zones of being. In every domain, the achievement of progress requires an excess of effort and therefore of force. Now whence does this increase come ?

In the abstract, one might assume an internal growth of the world's resources, an absolute increase in mechanical wealth corresponding to the expanding needs of evolution ; but, in fact, things seem to happen otherwise. In no case does the energy required for synthesis appear to be provided by an influx of fresh capital, but by expenditure. What is gained on one side is lost on the other. Nothing is constructed except at the price of an equivalent destruction.

Experimentally and at first sight, when we consider the universe in its mechanical functions, it does not reveal itself to us as an open quantum capable of containing an ever greater reality within its embrace, but as a closed quantum, within which nothing progresses except by exchange of that which was given in the beginning.

That is a first appearance.

*Second Principle.* In every physico-chemical change, adds thermodynamics, a fraction of the available energy is irrecoverably ' entropised ', lost, that is to say, in the form of heat. Doubtless it is possible to retain this degraded fraction symbolically in equations, so as to express that in the operations of matter nothing is lost any more than anything is created, but that is merely a mathematical trick. As a matter of fact, from the real evolutionary standpoint, something is finally burned in the course of every synthesis in order to pay for that synthesis. The more the energy-quantum of the world comes into play, the more it is consumed. Within the scope of our experience, the material concrete universe seems to be unable to continue on its way indefinitely in a closed cycle, but traces out irreversibly a curve of obviously limited development. And thus it is that this universe differentiates itself from purely abstract magnitudes and places itself among the realities which are born, which grow, and which die. From time it passes into duration ; and finally escapes from geometry dramatically to become, in its totality as in its parts, an object of history.[1]

Let us translate into images the natural significance of these two principles of the Conservation and Dissipation of Energy.

We said above that qualitatively the evolution of matter reveals itself to us, *hic et nunc*, as a process during which the constituents of the atom are inter-combined and ultra-condensed. Quantitatively, this transformation now appears to us as a definite, but costly, operation in which an original impetus slowly becomes exhausted. Laboriously, step by step, the atomic and molecular structures become higher and more complex, but the upward force is lost on the way. Moreover, the same wearing away that is gradually consuming the cosmos in its totality is at work within the terms of the synthesis, and the higher the terms the quicker this action takes place. Little by little, the *improbable* combinations that they represent become broken down again into more simple components, which fall back and are disaggregated in the shapelessness of *probable* distributions.

A rocket rising in the wake of time's arrow, that only bursts to be extinguished ; an eddy rising on the bosom of a descending current—such then must be our picture of the world.

So says science : and I believe in science : but up to now has science ever troubled to look at the world other than from *without* ?

[1] [cf. concluding sections of R. G. Collingwood : *Idea of Nature* (O.U.P. 1944).]

CHAPTER TWO

# THE WITHIN OF THINGS

On the scientific plane, the quarrel between materialists and the upholders of a spiritual interpretation, between finalists and determinists, still endures. After a century of disputation each side remains in its original position and gives its adversaries solid reasons for remaining there.

So far as I understand the struggle, in which I have found myself involved, it seems to me that its prolongation depends less on the difficulty that the human mind finds in reconciling certain apparent contradictions in nature—such as mechanism and liberty, or death and immortality—as in the difficulty experienced by two schools of thought in finding a common ground. On the one hand the materialists insist on talking about objects as though they only consisted of external actions in transient relationships. On the other hand the upholders of a spiritual interpretation are obstinately determined not to go outside a kind of solitary introspection in which things are only looked upon as being shut in upon themselves in their 'immanent' workings. Both fight on different planes and do not meet ; each only sees half the problem.

I am convinced that the two points of view require to be brought into union, and that they soon will unite in a kind of phenomenology or generalised physic in which the internal aspect of things as well as the external aspect of the world will be taken into account. Otherwise, so it seems to me, it is impossible to cover the totality of the cosmic phenomenon by one coherent explanation such as science must try to construct.

We have just described the *without* of matter in its connections and its measurable dimensions. Now, in order to advance still farther in the direction of man, we must extend the bases of our future edifices into the *within* of that same matter.

Things have their *within* ; their 'reserve', one might say ; and this appears to stand in definite *qualitative* or *quantitative* connections with the developments that science recognises in the cosmic energy. These three statements [i.e., that there is a *within*, that some connections are *qualitative*, that others are *quantitative*] are the basis of the three sections of this new chapter. To deal with them, as here I must, obliges me to overlap 'Before Life' and somewhat to anticipate 'Life' and 'Thought'. However, is not the peculiar difficulty of every synthesis that its end is already implicit in its beginnings ?

## 1. EXISTENCE

If there is one thing that has been clearly brought out by the latest advances in physics, it is that in our experience there are 'spheres' or 'levels' of different kinds in the unity of nature, each of them distinguished by the dominance of certain factors which are imperceptible or negligible in a neighbouring sphere or on an adjacent level. On the middle scale of our organisms and of our constructions velocity does not seem to change the nature of matter. None the less, we now know that at the extreme values reached by atomic movements it profoundly modifies the mass of bodies. Among 'normal' chemical elements, stability and longevity appear to be the rule : but that illusion has been destroyed by the discovery of radio-active substances. By the standards of our human existence, the mountains and stars are a model of majestic changelessness. Now we discover that, observed over a sufficiently great duration of time, the earth's crust changes ceaselessly under our feet, while the heavens sweep us along in a cyclone of stars.

In all these instances, and in others like to them, there is no absolute appearance of a new dimension. *Every* mass is modified by its velocity. *Every* body radiates. *Every* movement is veiled in immobility when sufficiently slowed down. But on a different scale, or at a different intensity, there will become visible some phenomenon that spreads over the horizon, blots out the other distinctions, and gives its own particular tonality to the whole picture.

It is the same with the *within* of things.

For a reason that will soon appear, objects in the realm of physico-chemistry are only made manifest by their outward determinisms.

In the eyes of the physicist, nothing exists legitimately, at least up to now, except the *without* of things. The same intellectual attitude is still permissible in the bacteriologist, whose cultures (apart from some substantial difficulties) are treated as laboratory reagents. But it is already more difficult in the realm of plants. It tends to become a gamble in the case of a biologist studying the behaviour of insects or coelenterates. It seems merely futile with regard to the vertebrates. Finally, it breaks down completely with man, in whom the existence of a *within* can no longer be evaded, because it is the object of a direct intuition and the substance of all knowledge.

The apparent restriction of the phenomenon of consciousness to the higher forms of life has long served science as an excuse for eliminating it from its models of the universe. A queer exception, an aberrant function, an epiphenomenon—thought was classed under one or other of these heads in order to get rid of it. But what would have happened to modern physics if radium had been classified as an 'abnormal substance' without further ado? Clearly, the activity of radium had not been neglected, and could not be neglected, because, being measurable, it forced its way into the external web of matter—whereas consciousness, in order to be integrated into a world-system, necessitates consideration of the existence of a new aspect or dimension in the stuff of the universe. We shrink from the attempt, but which of us does not in both cases see an identical problem facing research workers, which have to be solved by the same method, namely, *to discover the universal hidden beneath the exceptional* ?

Latterly we have experienced it too often to admit of any further doubt: an irregularity in nature is only the sharp exacerbation, to the point of perceptible disclosure, of a property of things diffused throughout the universe, in a state which eludes our recognition of its presence. Properly observed, even if only in one spot, a phenomenon necessarily has an omnipresent value and roots by reason of the fundamental unity of the world. Whither does this rule lead us if we apply it to the instance of human 'self-knowledge' ?

'Consciousness is completely evident only in man' we are tempted to say, 'therefore it is an isolated instance of no interest to science.'

'Consciousness is evident in man,' we must continue, correcting ourselves, 'therefore, half-seen in this one flash of light, it has a cosmic extension, and as such is surrounded by an aura of indefinite spatial and temporal extensions.'

The conclusion is pregnant with consequences, and yet I cannot see how, by sound analogy with all the rest of science, we can escape from it.

It is impossible to deny that, deep within ourselves, an 'interior' appears at the heart of beings, as it were seen through a rent. This is enough to ensure that, in one degree or another, this 'interior' should obtrude itself as existing everywhere in nature from all time. Since the stuff of the universe has an inner aspect at one point of itself, there is necessarily a *double aspect to its structure*, that is to say in every region of space and time—in the same way, for instance, as it is granular: *co-extensive with their Without, there is a Within to things.*

The consequent picture of the world daunts our imagination, but it is in fact the only one acceptable to our reason. Taken at its lowest point, exactly where we put ourselves at the beginning of these pages, primitive matter is something more than the particulate swarming so marvellously analysed by modern physics. Beneath this mechanical layer we must think of a 'biological' layer that is attenuated to the uttermost, but yet is absolutely necessary to explain the cosmos in succeeding ages. The *within*, *consciousness*[1] and then *spontaneity*—three expressions for the same thing. It is no more legitimate for us experimentally to fix an absolute beginning to these three expressions of one and the same thing than to any other lines of the universe.

*In a coherent perspective of the world: life inevitably assumes a 'pre-life' for as far back before it as the eye can see.*[2]

In that case—and the objection will come from materialists and upholders of spirituality alike—if everything in nature is basically living, or at least pre-living, how is it possible for a mechanistic science of matter to be built up and to triumph?

Determinate *without*, and 'free' *within*—would the two aspects of things be irreducible and incommensurable? If so, where is your solution?

The answer to this difficulty is already implicit in what we

[1] Here, and throughout this book, the term 'consciousness' is taken in its widest sense to indicate every kind of psychism from the most rudimentary forms of interior perception imaginable to the human phenomenon of reflective thought.

[2] These pages had been written for some time when I was surprised to find their substance in some masterly lines recently written by J. B. S. Haldane:

'We do not find obvious evidence of life or mind in so-called inert matter, and we naturally study them most easily where they are most completely manifested; but if the scientific point of view is correct, we shall ultimately find them, at least in rudimentary forms, all through the universe.'

And he goes on to add these words which my readers would do well to recall when I come to unveil (with all due reservations and corrections) the perspective of the 'Omega Point':

'Now, if the co-operation of some thousands of millions of cells in our brain can produce our consciousness, the idea becomes vastly more plausible that the co-operation of humanity, or some sections of it, may determine what Comte calls a Great Being.' (Essay on Science and Ethics in *The Inequality of Man*, Chatto, 1932, p. 113.)

What I say is thus not absurd. Moreover, any metaphysician must rejoice to discover that even in the eyes of physics the idea of absolutely brute matter (that is to say, of a pure 'transient') is only a first very rough approximation of our experience.

have said above about the diversity of 'spheres of experience' superposed in the interior of the world. It will appear more clearly when we have discerned the qualitative laws that govern in their growth and variation the manifestations of what we have just called the *within* of things.

## 2. THE QUALITATIVE LAWS OF GROWTH

To harmonise objects in time and space, without presuming to determine the conditions that can rule their deepest being : to establish an experimental chain of succession in nature, not a union of 'ontological' causality ; to see, in other words, and not to explain—this, let it not be forgotten, is the sole aim of the present study.

From this phenomenal point of view (which is *the* scientific point of view) can one go beyond the position where our analysis of the stuff of the universe has just stopped ? In this last we have recognised the existence of a conscious inner face that everywhere duplicates the 'material' external face, which alone is commonly considered by science. Can we go further and define the rules according to which this second face, for the most part entirely hidden, suddenly shows itself, and then as suddenly bursts through into certain other regions of our experience ?

Yes, so it seems, and even quite easily, provided there are placed one after the other three observations that each one of us could have made, but which do not take on their true value until we think of linking them together.

### A. *First Observation*

Considered in its pre-vital state, the *within* of things, whose reality even in the nascent forms of matter we have just admitted, must not be thought of as forming a continuous film, but as assuming the same granulation as matter itself.

Soon we shall have to return to this essential point. As far back as we began to descry them, *the first living things* reveal themselves to our experience as kinds of 'mega-' or 'ultra-' molecules, both in size and in number: a bewildering multitude of microscopic nuclei. Which means that for reasons of homogeneity and continuity, the pre-living can be divined, below the horizon, as an object sharing in the *corpuscular* structure and properties of the world. Looked at from *within*, as well as observed from *without*, the stuff of the universe thus tends likewise to be resolved backwardly into a dust of particles that are (i) perfectly alike among themselves (at least if they are observed from a great distance); (ii) each co-extensive with the whole of the cosmic realm; (iii) mysteriously connected among themselves, finally, by a global energy. In these depths the world's two aspects, external and internal, correspond point by point. So much is this so that one may pass from the one to the other on the sole condition that 'mechanical interaction' in the definition of the partial centres of the universe given above is replaced by 'consciousness'.

*Atomicity is a common property of the Within and the Without of things.*

### B. *Second Observation*

Virtually homogeneous among themselves in the beginning, the elements of consciousness, exactly as the elements of matter which they subtend, complicate and differentiate their nature, little by little, with the passage of duration. From this point of view and considered solely from the experimental aspect, consciousness reveals itself as a cosmic property of variable size subject to a global transformation. Taken on the ascent, this huge phenomenon that we shall have to follow all along the development of life right up to the appearance of thought, has ended by appearing commonplace. Followed in the opposite direction, it leads us, as we have already seen, to the less familiar idea of inferior states that are ever less well defined and, as it were, distended.

*Refracted rearwards along the course of evolution, consciousness displays itself qualitatively as a spectrum of shifting shades whose lower terms are lost in the night.*

### c. *Third Observation*

Finally, let us take from two different regions of this spectrum two particles of consciousness that are at unlike stages of evolution. As we have seen, there corresponds to each of them, by construction, a certain definite material grouping of which they form the *within*. Let us compare these two external groupings the one with the other and ask ourselves how they are arranged with regard to each other and with regard to the portion of consciousness that each of them encloses.

The answer comes at once.

Whatever instance we may think of, we may be sure that every time a richer and better organised structure will correspond to the more developed consciousness.

The simplest form of protoplasm is already a substance of unheard-of complexity. This complexity increases in geometrical progression as we pass from the protozoon higher and higher up the scale of the metazoa. And so it is for all the rest always and everywhere. Here again, the phenomenon is so obvious that we have long since ceased to be astonished by it. Yet its importance is decisive. For thanks to it we possess a tangible 'parameter' allowing us to connect both the internal and the external films of the world, not only *in their position* (point by point), but also, as we shall verify later on, *in their motion*.

The degree of concentration of a consciousness varies in inverse ratio to the *simplicity* of the material compound lined by it. Or again : a consciousness is that much more perfected according as it lines a richer and better organised material edifice.

*Spiritual perfection* (*or conscious* '*centreity*') *and material syn-*

*thesis (or complexity) are but the two aspects or connected parts of one and the same phenomenon.*[1]

And now we have arrived, *ipso facto*, at the solution of the problem posed for us. We are seeking a qualitative law of development that from sphere to sphere should be capable of explaining, first of all the invisibility, then the appearance, and then the gradual dominance of the *within* in comparison to the *without* of things. This law reveals itself once the universe is thought of as passing from *State A*, characterised by a very large number of very simple material elements (that is to say, with a very poor *within*), to *State B* defined by a smaller number of very complex groupings (that is to say, with a much richer *within*).

In State A, the centres of consciousness, because they are extremely numerous and extremely loose at the same time, only reveal themselves by overall effects which are subject to the laws of statistics. Collectively, that is, they obey the laws of mathematics. This is the proper field of physico-chemistry.

In State B, on the other hand, these less numerous[2] and at the same time more highly individualised elements gradually escape from the slavery of large numbers. They allow their basic non-measurable spontaneity to break through and reveal itself. We can begin to see them and follow them one by one, and in so doing we have access to the world of biology.

In sum, all the rest of this essay will be nothing but the story of the struggle in the universe between the unified *multiple* and the unorganised *multitude* : the application throughout of the great *Law of complexity and consciousness* : a law that itself implies a psychically convergent structure and curvature of the world.

But we must not go too quickly, and since we are still con-

[1] From this aspect one might say that, on the phenomenal plane, each being is constructed like an ellipse on two conjugate foci : a focus of material organisation and a focus of psychic centering—the two foci varying solidarily and in the same sense.

[2] As we shall see, this is despite the specifically vital mechanism of *multiplication*.

cerned with pre-life let us only keep in mind that, from the *qualitative* viewpoint, there is no kind of contradiction involved in admitting that a universe of mechanistic appearance may be built up of 'liberties'—provided that the liberties are therein contained in a sufficiently fine state of division and imperfection.

## 3. SPIRITUAL ENERGY

There is no concept more familiar to us than that of spiritual energy, yet there is none that is more opaque scientifically. On the one hand the objective reality of psychical effort and work is so well established that the whole of ethics rests on it and, on the other hand, the nature of this inner power is so intangible that the whole description of the universe in mechanical terms has had no need to take account of it, but has been successfully completed in deliberate disregard of its reality.

The difficulties we still encounter in trying to hold together spirit and matter in a reasonable perspective are nowhere more harshly revealed. Nowhere either is the need more urgent of building a bridge between the two banks of our existence—the physical and the moral—if we wish the material and spiritual sides of our activities to be mutually enlivened.

To connect the two energies, of the body and the soul, in a coherent manner: science has provisionally decided to ignore the question, and it would be very convenient for us to do the same. Unfortunately, or fortunately, caught up as we are here in the logic of a system where the *within* of things has just as much or even more value than their *without*, we collide with the difficulty head on. It is impossible to avoid the clash: we must advance.

Naturally the following considerations do not pretend to be a truly satisfactory solution of the problem of spiritual energy. Their aim is merely to show by means of one example what, in my opinion, an integral science of nature should adopt as its line of research and the kind of interpretation it should follow.

### A. *The Problem of the Two Energies*

Since the inner face of the world is manifest deep within our human consciousness, and there reflects upon itself, it would seem that we have only got to look at ourselves in order to understand the dynamic relationships existing between the *within* and the *without* of things at a given point in the universe.

In fact so to do is one of the most difficult of all things.

We are perfectly well aware in our concrete actions that the two opposite forces combine. The motor works, but we cannot make out the method, which seems to be contradictory. What makes the crux—and an irritating one at that—of the problem of spiritual energy for our reason is the heightened sense that we bear without ceasing in ourselves that our action seems at once to depend on, and yet to be independent of, material forces.

First of all, the dependence. This is depressingly and magnificently obvious. 'To think, we must eat.' That blunt statement expresses a whole economy, and reveals, according to the way we look at it, either the tyranny of matter or its spiritual power. The loftiest speculation, the most burning love are, as we know only too well, accompanied and paid for by an expenditure of physical energy. Sometimes we need bread, sometimes wine, sometimes a drug or a hormone injection, sometimes the stimulation of a colour, sometimes the magic of a sound which goes in at our ears as a vibration and reaches our brains in the form of inspiration.

Without the slightest doubt *there is something* through which material and spiritual energy hold together and are complementary. In last analysis, *somehow or other*, there must be a single energy operating in the world. And the first idea that occurs to us is that the 'soul' must be as it were a focal point of transformation at which, from all the points of nature, the forces of bodies converge, to become interiorised and sublimated in beauty and truth.

Yet, seductive though it be, the idea of the *direct* transformation of one of these two energies into the other is no sooner glimpsed than it has to be abandoned. As soon as we try to couple them together, their mutual independence becomes as clear as their interrelation.

Once again : ' To think, we must eat.' But what a variety of thoughts we get out of one slice of bread ! Like the letters of the alphabet, which can equally well be assembled into nonsense as into the most beautiful poem, the same calories seem as indifferent as they are necessary to the spiritual values they nourish.

The two energies—of mind and matter—spread respectively through the two layers of the world (the *within* and the *without*) have, taken as a whole, much the same demeanour. They are constantly associated and in some way pass into each other. But it seems impossible to establish a simple correspondence between their curves. On the one hand, only a minute fraction of ' physical ' energy is used up in the highest exercise of spiritual energy ; on the other, this minute fraction, once absorbed, results on the internal scale in the most extraordinary oscillations.

A quantitative disproportion of this kind is enough to make us reject the naïve notion of ' change of form ' (or direct transformation)—and hence all hope of discovering a ' mechanical equivalent ' for will or thought. Between the *within* and the *without* of things, the interdependence of energy is incontestable. But it can in all probability only be expressed by a complex symbolism in which terms of a different order are employed.

### B. *A Line of Solution*

To avoid a fundamental dualism, at once impossible and anti-scientific, and at the same time to safeguard the natural complexity of the stuff of the universe, I accordingly propose the following as a basis for all that is to emerge later.

We shall assume that, essentially, all energy is psychic in nature ; but add that in each particular element this fundamental energy is divided into two distinct components : a *tangential energy* which links the element with all others of the same order (that is to say, of the same complexity and the same centricity) as itself in the universe ; and a *radial energy* which draws it towards ever greater complexity and centricity—in other words forwards.[1]

From this initial state, and supposing that it disposes of a certain free tangential energy, the particle thus constituted must obviously be in a position to increase its internal complexity in association with neighbouring particles, and thereupon (since its centricity is automatically increased) to augment its radial energy. The latter will then be able to react in its turn in the form of a new arrangement in the tangential field. And so on.

In this view, whereby tangential energy represents 'energy' as such, as generally understood by science, the only difficulty is to explain the interplay of tangential arrangements in terms of the laws of thermo-dynamics. As regards this we may remark the following :

*a.* First of all, since the variation of radial energy in function of tangential energy is effected, according to our hypothesis, *by the intervention of an arrangement*, it follows that as much as you like of the first may be linked with as little as you like of the second—for a highly perfected arrangement may only require an extremely small amount of work. This fits in with the facts noted in section A above.

*b.* Moreover, in the system here proposed, we are paradoxically led to admit that cosmic energy is constantly increasing, not only in its radial form, but—which is much more serious—in its tangential one (for the tension between elements increases with

[1] Let it be noted in passing that the less an element is 'centred' (i.e. the feebler its radial energy) the more will its tangential energy reveal itself in powerful mechanical effects. Between strongly 'centred' particles (i.e. of high radial energy) the tangential seems to become 'interiorised' and to disappear from the physicist's view. Probably we have here an auxiliary principle which could help to explain the apparent conservation of energy in the universe (see para. *b.* below). We probably ought to recognise *two* sorts of tangential energy, one of *radiation* (at its maximum with the lowest radial values, as in the atom), the other of *arrangement* (only appreciable with the highest radial values, as in living creatures, man in particular).

their centricity itself). This would seem to be in direct contradiction with the law of conservation of energy. It must be noted, however, that this increase of the tangential of the second kind (the only one troublesome for physics) only becomes appreciable with very high radial values (as in man, for instance, and social tensions). Below this level, and for an approximately constant number of initial particles in the universe, the sum of the cosmic tangential energies remains practically and statistically invariable in the course of transformations. And this is all that science requires.

*c.* Lastly, since according to our reading, the entire edifice of the universe is constantly supported at every phase of its progressive ' centration ' by its primary arrangements, it is plain that its achievement will be conditioned up to the highest stages by a certain primordial quantum of free tangential energy, which will gradually exhaust itself, following the principle of entropy.

Looked at as a whole, this picture satisfies the requirements of reality.

Three questions remain still unanswered, however :

*a.* By virtue of what special energy does the universe propagate itself along its main axis in the less probable direction of the higher forms of complexity and centricity ?

*b.* Is there a definite limit and end to the ' elemental ' value and to the sum total of the radial energies developed in the course of transformation ?

*c.* Is this final and resultant form of radial energies, supposing it exists, subject to reversal? Is it destined one day to start disintegrating so as to satisfy the principle of entropy, and fall back indefinitely into pre-living and still lower centres, by the exhaustion and gradual levelling-down of the free tangential energy contained in the successive envelopes of the universe from which it has emerged ?

To be answered satisfactorily, these three questions must await a much later chapter, when the study of man will have led us to the concept of a superior pole to the world—the *omega point.*

CHAPTER THREE

# THE EARTH IN ITS EARLY STAGES

SOME THOUSANDS of millions of years ago, not, it would appear, by a regular process of astral evolution, but as the result of some unbelievable accident (a brush with another star? an internal upheaval ?) a fragment of matter composed of particularly stable atoms was detached from the surface of the sun. Without breaking the bonds attaching it to the rest, and just at the right distance from the mother-star to receive a moderate radiation, this fragment began to condense, to roll itself up, to take shape.[1] Containing within its globe and orbit the future of man, another heavenly body—a planet this time—had been born.

So far our eyes have been straying over the unlimited layers in which the stuff of the universe is deployed.

From now on let us concentrate our attention on this diminutive, obscure, but fascinating object which had just appeared. *It is the only place* in the world in which we are so far able to study the evolution of matter in its ultimate phases, and as far as ourselves.

Let us have a look at the earth in its early stages, so fresh yet charged with latent powers, as it balances in the chasms of the past.

[1] Once again astronomers seem to be returning to a more Laplacean concept of the birth of planets by the effect of knots and bulges in the cloud of cosmic dust originally floating round each star.

## 1. THE WITHOUT

What arouses the physicist's interest in this globe—new-born, it would seem, by a stroke of chance in the cosmic mass—is the presence of composite chemical bodies not to be observed anywhere else.[1] At the extreme temperature occurring in the stars, matter can only survive in its most dissociated states. Only simple bodies exist on these incandescent stars. On the earth this simplicity of the elements still obtains at the periphery, in the more or less ionised gases of the atmosphere and the stratosphere and, probably, far below, in the metals of the 'barysphere'. But between these two extremes comes a long series of complex substances, harboured and produced only by stars that have 'gone out'. Arranged in successive zones, they demonstrate from the start the powers of synthesis contained in the universe. First the siliceous zone, preparing the solid crust of the planet. Next the zone of water and carbonic acid, enclosing the silicates in an unstable, mobile and penetrating envelope.

In other words we have the barysphere, lithosphere, hydrosphere, atmosphere and stratosphere.

This fundamental composition may have varied and become elaborated in detail, but by and large it can be said to have established itself from the beginning. And it is from it that geochemistry develops progressively in two different directions.

### A. *The Crystallising World*

In one direction, much the more common, terrestrial energy has tended from the outset to be given off and liberated. Silicates, water, carbon dioxide—these essential oxides were formed by burning up and neutralising (alone or in association with other simple bodies) the affinities of their elements. Carrying the

[1] Except, though very fugitively, in the atmosphere of the planets nearest to our own.

scheme progressively further, the result is the rich variety of the 'mineral world'.

The mineral world is a much more supple and mobile world than could be imagined by the science of the ancients. Vaguely analogous to the metamorphoses of living creatures, there occurs in the most solid rocks, as we now know, perpetual transformation of a mineral species.

But it is a world relatively poor in compounds, because of the narrow limit to the internal architecture of its elements. According to latest estimates, we have found only a few hundred silicates in nature.

Looking at them 'biologically' we may say it is the characteristic of minerals (as of so many other organisms that have become incurably fixed) to have chosen a road which closed them prematurely in upon themselves. By their innate structure the molecules are unfitted for growth. To develop beyond a certain size they have in a way to get out of themselves, to have recourse to a trick of purely external association, whereby the atoms are linked together without true combination or union. Sometimes we find them in strings as in jade, sometimes in planes as in mica, and sometimes in a solid quincunx as in garnet.

In this way, by simple juxtaposition of atoms or relatively simple atomic groups in geometrical patterns, regular aggregates may be produced whose level of composition is often very high, but they correspond to no properly centred units ; they are an indefinitely extended mosaic of small elements—such as we know to be the structure of a crystal, which, thanks to X-rays, can now be photographed. And such is the organisation, simple and stable, which the condensed matter around us has by and large perforce adopted from its origins.

Considered in the mass, the earth is veiled in geometry as far back as we can see. It crystallises.

But not completely.

### B. *The Polymerising World*

In the course of and by virtue of the initial advance of the elements on earth towards the crystalline state, energy was constantly released and liberated (just as, today, it is released by mankind as a result of machinery). This was constantly augmented by energy furnished by the atomic decomposition of radio-active substances and by that given off by solar rays. Where could this surplus energy, available on the surface of the earth in its early stages, go to ? Was it merely to be lost around the globe in obscure emanations ?

Another much more probable hypothesis occurs to us when we look at the world today. When it became too weak to escape in incandescence, the free energy of the new-born earth became capable of reacting on itself in a work of synthesis. Thus, as today, it passed with the absorption of heat into building up certain carbonates, hydrates or hydrites, and nitrates like those which astonish us by their power to increase indefinitely the complexity and instability of their elements. This is the realm of *polymerisation*,[1] in which the particles ' concatenate ', group themselves and exchange positions, as in crystals, in a theoretically endless network. *Only, this time it is molecules with molecules in such a way as to form on each occasion (by closed or at all events limited combination) an ever larger and more complex molecule.*

This world of ' organic compounds ' is ours. We live among them and are made of them. So intimately do we see it as connected with the phenomena of life that we have got into the habit of considering it only in direct association with life *already constituted*. Moreover, despite its incredible wealth of forms, which far surpasses the variety of mineral compounds, it concerns such a tiny part of the substance of the earth that we are instinctively

[1] I trust I shall be forgiven (as later in the case of ' orthogenesis ') for using this term in so generalised a sense, i.e. to include (as well as the strict polymerisation of the chemists) the entire process of ' additive complexification ' producing large molecules.

inclined to relegate it to a minor position of geo-chemistry—like the ammonia and oxides that surround the lightning's flash.

If we wish later to fix the place of man in nature, it seems to me essential to restore to this phenomenon its true physiognomy and its 'seniority'.

Whatever the quantitative disproportion of the masses they respectively involve, inorganic and organic chemistry are only and can only be two inseparable facets of one and the same telluric operation. And the second, no less than the first, must be regarded as already under way in the infancy of the earth. We are back at the refrain that runs all the way through this book. *In the world, nothing could ever burst forth as final across the different thresholds successively traversed by evolution (however critical they be) which has not already existed in an obscure and primordial way.* If the organic had not existed on earth from the first moment at which it was possible, it would never have begun later.

There is good reason to think that around our nascent planet, in addition to the inchoation of a metallic barysphere, a siliceous lithosphere, a hydrosphere and an atmosphere, there was the outline of a special envelope, the antithesis, we might say, of the first four: the temperate zone of polymerisation, in which water, ammonia and carbon dioxide were already floating in the rays of the sun. To ignore that tenuous film would be to deprive the infant earth of its most essential adornment. For, as we shall see, it is in this that the '*within* of the earth' was soon to be gradually concentrated (if we hold to what I have already said).

## 2. THE WITHIN

When I speak of the '*within*' of the earth, I do not of course mean those material depths in which—a few miles beneath our feet—lurks one of the most vexatious mysteries of science: the chemical nature and the exact physical condition of the internal regions of the globe. The '*within*' is used here, as in the preceding chapter, to denote the 'psychic' face of that portion of the stuff of the cosmos enclosed from the beginning of time within the narrow scope of the early earth. In that fragment of sidereal matter which has just been isolated, as in every other part of the universe, the exterior world must inevitably be lined at every point with an interior one. This we have shown already. Only here the conditions have changed. Matter no longer spreads out beneath our eyes in diffuse and undefinable layers. It coils up round itself in a *closed volume. How will its 'inner' layer react to such involution?*

First let it be noted that, by the very fact of the individualisation of our planet, a certain mass of elementary consciousness was originally emprisoned in the matter of earth. Some scientists have felt obliged to invest some interstellar germs with the power of fecundating cooling stars. This hypothesis disfigures, without explaining, the wonderful phenomenon of life, with its noble corollary, the phenomenon of man. It is in fact quite useless. Why should we turn to space to look for a fecundating principle for the earth—which is incomprehensible in any case? By its initial chemical composition, the early earth is itself, and in its totality, the incredibly complex germ we are seeking. Congenitally, if I may use the word, it already carried pre-life within it, and this, moreover, in *definite quantity*. The whole question is to define how, from this primitive and essentially elastic quantum, all the rest has emerged.

To form an idea of the first phases of this evolution it will be enough to compare, stage by stage, on the one hand the general laws we have felt able to lay down for the development of spiritual energy, and on the other the physico-chemical conditions we have just acknowledged in the nascent earth. We have said that spiritual energy, by its very nature, increases in 'radial' value, positively, absolutely, and without determinable limits, in step with the increasing chemical complexity of the elements of which it represents the inner lining. But the chemical complexity of the earth increases in conformity with the laws of thermo-dynamics in the particular, superficial zone in which its elements polymerise.

If we put these two propositions side by side we see that they interweave and shed light upon each other without ambiguity. With one accord they tell us that pre-life is no sooner enclosed in the nascent earth than it emerges from the torpor to which it appeared to have been condemned by its diffusion in space. Its activities, hitherto dormant, are now set in motion *pari passu* with the awakening of the forces of synthesis enclosed in matter. And at one and the same stroke, over the whole surface of the new-formed globe, the tension of internal freedoms begins to rise.

Let us look more attentively at this mysterious surface.

A character to be noted at the outset is the extremely small size and the extremely great number of the particles of which it consists. For a thickness of some miles, in water, in air, in muddy deposits, ultra-microscopic grains of protein are thickly strewn over the surface of the earth. Our imaginations boggle at the mere thought of counting the flakes of this snow. Yet if we take it that pre-life has already emerged in the atom, are not these myriads of large molecules just what we ought to expect?

But there is another point to consider.

In a sense more remarkable than their multitude (and as important to keep in mind for future developments) is the solidarity due to their very genesis which unites the specks of this primordial dust of consciousness. That which permits the growth of elementary freedoms is, essentially, I repeat, the growing synthesis of the molecules they subtend. And let me also repeat that this synthesis itself would never take place if the globe as a whole did not enfold within a closed surface the layers of its substance.

Thus, wherever we look on earth, the growth of the '*within*' only takes place thanks to a *double related involution*, the coiling up of the molecule upon itself and the coiling up of the planet upon itself.[1] The initial quantum of consciousness contained in our terrestrial world is not formed merely of an aggregate of particles

[1] Precisely the conditions we find later on, at the other end of evolution, presiding over the genesis of the '*noosphere*'.

caught fortuitously in the same net. It represents a correlated mass of infinitesimal centres structurally bound together by the conditions of their origin and development.

Here again, but in a better defined field and on a higher level, we find the fundamental condition characteristic of primordial matter—the unity of plurality. The earth was probably born by accident ; but, in accordance with one of the most general laws of evolution, scarcely had this accident happened than it was immediately made use of and recast into something naturally directed. By the very mechanism of its birth, the film in which the ' *within* ' of the earth was concentrated and deepened emerges under our eyes in the form of an organic whole in which no element can any longer be separated from those surrounding it. Another ' indivisible ' has appeared at the heart of the great ' indivisible ' which is the universe. In truth, a *pre-biosphere*.

And this is the envelope which, taken in its entirety, is to be our sole preoccupation from now on.

As we continue peering into the abysses of the past, we can see its colour changing.

From age to age it increases in intensity. Something is going to burst out upon the early earth, and this thing is Life.

BOOK TWO

# LIFE

CHAPTER ONE

# THE ADVENT OF LIFE

AFTER WHAT we have said about the latent germinal powers of the early earth, it might be thought that nothing had been left in nature which could pin-point the beginning of life, and that therefore my chapter heading is inappropriate. The mineral world and the world of life seem two antithetical creations when viewed by a summary glance in their extreme forms and on the intermediary scale of our human organisms ; but to a deeper study, when we force our way right down to the microscopic level and beyond to the infinitesimal, or (which comes to the same thing) far back along the scale of time, they seem quite otherwise—a single mass gradually melting in on itself.

At such depths all differences seem to become tenuous. For a long time we have known how impossible it is to draw a clear line between animal and plant on the unicellular level. Nor can we draw one (as we shall see later) between 'living' protoplasm and 'dead' proteins on the level of the very big molecular accumulations. We still use the word 'dead' for these latter unclassified substances, but have we not already come to the conclusion that they would be incomprehensible if they did not possess already, deep down in themselves, some sort of rudimentary psyche ?

So, in a sense, we can no more fix an absolute zero in time (as was once supposed) for the advent of life than for that of any other experimental reality. On the experimental and phenomenological plane, a given universe and each of its parts can only have one and the same duration, to which there is no backward limit.

Thus each thing extends itself and pushes its roots into the past, ever farther back, by that which makes it most itself. Everything, in some extremely attenuated extension of itself, has existed from the very first. Nothing can be done in a direct way to counter this basic condition of our knowledge.

But to have realised and accepted once and for all that each new being has and must have a *cosmic embryogenesis* in no way invalidates the reality of its *historic birth*.

In every domain, when anything exceeds a certain measurement, it suddenly changes its aspect, condition or nature. The curve doubles back, the surface contracts to a point, the solid disintegrates, the liquid boils, the germ cell divides, intuition suddenly bursts on the piled up facts . . . Critical points have been reached, rungs on the ladder, involving a change of state—jumps of all sorts *in the course* of development. Henceforward this is the *only* way in which science can speak of a ' first instant '. But it is none the less a *true* way.

In this new and more complicated sense—even after (precisely after) what we have said about pre-life—our task, now, is to consider and define a beginning of life.

Through a duration to which we can give no definite measure but know to be immense, the earth, cool enough now to allow the formation on its surface of the chains of molecules of the carbon type, was probably covered by a layer of water from which emerged the first traces of future continents. To an observer equipped with even the most modern instruments of research, our earth would probably have seemed an inanimate desert. Its waters would have left no trace of mobile particles even upon the finest of our filters, and the most powerful microscope would only have detected inert aggregates.

Then at a given moment, after a sufficient lapse of time, those same waters here and there must unquestionably have begun writhing with minute creatures. And from that initial proliferation stemmed the amazing profusion of organic matter whose matted complexity came to form the last (or rather the last but one) of the envelopes of our planet : the *biosphere*.

No amount of historical research will ever reveal the details of this story. Unless the science of tomorrow is able to reconstruct the process in the laboratory, we shall probably never find any material vestige of this emergence of the microscopic from the molecular, of the organic from the chemical, of the living from the pre-living. One thing is certain, however—a metamorphosis of this sort could not be the result of a simple continuous process. By analogy with all we have learnt from the comparative study of natural developments, we must postulate at this particular moment of terrestrial evolution a coming to maturity, a threshold, a crisis of the first magnitude, the beginning of a new order.

We shall now try to determine what must have been on the one hand the nature, on the other the spatial and temporal modalities of this transformation ; and find an explanation that will fit in both with what we presume to have been the conditions on the early earth and with those of the earth as it is today.

## 1. THE TRANSIT TO LIFE

Seen from outside and materially, the best we can say at the moment is that life properly speaking *begins with the cell.* For a century science has concentrated its attention on this chemically and structurally ultra-complex unit, and the longer it continues to do so the more evident it becomes that in it lies the secret of which we have as yet no more than an inkling—the secret of the connection between the two worlds of physics and biology. The cell is *the natural granule of life* in the same way as the atom is the natural granule of simple, elemental matter. If we are to take the measure of the transit to life and determine its precise nature, we must try to understand the cell.

But to understand it, how are we to regard it ?

Volumes have been written about the cell. Whole libraries are insufficient to contain all that has been meticulously observed concerning its texture, the functions of its 'cytoplasm' and nucleus, the way it divides, and its connection with heredity. Yet, in itself, it is still a closed book, still as enigmatic as ever. It seems as though, once we have reached a certain depth in our explanation, we find ourselves reduced to marking time in front of an impregnable fortress.

It might seem that the histological and physiological methods of analysis have given us all we could expect of them and that, to get any farther, our approach must be made from another angle.

For obvious reasons, cytology has so far proceeded with an almost exclusively biological outlook. The cell has been viewed as a micro-organism, or an example of proto-life, that must be interpreted in relation to its highest forms and associations.

But this attitude has left half our problem in the dark. Like the moon in its first quarter, the cell has been illumined only on the side that looks towards the highest forms of life, leaving the other side (the layers we have called pre-life) floating in darkness. That is most likely the reason scientifically speaking why its mystery has been so unduly prolonged.

Marvellous as it is, marvellous as it seems to us in its isolation among the other constructions of matter, the cell, like everything else in the world, cannot be *understood* (i.e. incorporated in a coherent system of the universe) unless we situate it on an evolutionary line between a past and a future. We have turned a good deal of attention to its development and its differentiations. It is on its origins, that is to say on its roots in the inorganic, that we must now focus our researches if we want to grasp the essence of its novelty.

Despite what experience has taught us in every other field, we have let ourselves become too much accustomed to thinking of the cell as an object without antecedents. Let us see what happens if we regard it and treat it (as we certainly should) as something *at one and the same time* both the outcome of long preparation and yet profoundly original, that is to say, as a thing that is born.

### A. *Micro-organisms and Mega-molecules*

First of all the preparatory process.

When we try to look at the beginning of life in relation to its antecedents rather than its consequents, we at once notice something which, strangely enough, had never struck us before. It is in and by means of the cell that the molecular world 'appears in person' (if I may so express myself), touching, passing into, and disappearing in the higher constructions of life.

Perhaps a word of explanation is needed.

When we look at bacteria, it is always against a background of the higher plants and animals, and this blinds our vision. What we should do is start from another angle, shutting our eyes to all the more advanced forms in living nature and even to most of the protozoa because, in their main lines, they are almost as differentiated as metazoa. In the latter, moreover, let us ignore the highly specialised and often very large cells of the nervous, muscular and reproductive systems. In other words let us confine ourselves to the more or less independent elements, externally amorphous or polymorphous, such as abound in natural ferments, are present in our blood and accumulate in our organs in the form of connective tissue, in other words let us confine ourselves to what appear the simplest and the most primitive cells in nature today. This done, let us look at this corpuscular mass in relation to the matter beneath it. Can we fail for a moment to see the obvious relationship, in both composition and appearance, between the proto-living world on the one hand and the physico-chemical one on the other? When we consider the simplicity of the cellular form, the structural symmetry, the infinitesimal size, the outer uniformity in character and behaviour in the mass or multitude, do we not find the unmistakable characteristics and habits of the granular formations? In other words, we are still on that first rung of life, if not at the heart of 'matter', at least fully on its border.

Without exaggeration it may be said that just as man, seen in terms of palaeontology, merges anatomically with the mass of mammals that preceded him, so, *probing backwards*, we see the cell merging qualitatively and quantitatively with the world of chemical structures. Followed in a backward direction, it visibly converges towards the molecule.

This is already something more than a simple intellectual intuition.

Only a few years ago what I have just said concerning the gradual conversion of the 'granule' of matter into the 'granule' of life might have been thought of as being as suggestive, but at the same time as unfounded, as the first dissertations of Darwin or Lamarck on evolution. But things are now changing. Since the days of Darwin and Lamarck, numerous discoveries have established the existence of the transitional forms postulated by the theory of evolution. At the same time the latest advances in biochemistry are beginning to establish the reality of molecular aggregates which really do appear to reduce to measurable proportions the gaping void hitherto supposed to exist between protoplasm and mineral matter. If certain calculations (admittedly indirect) are accepted as correct, the molecular weights of some of the natural proteinous substances (such as the viruses so mysteriously associated with the zymotic diseases in plants and animals) may well be *in terms of millions*. Much smaller than any bacteria—so small in fact that no filter can retain them—the particles forming these substances are none the less colossal compared with the molecules normally dealt with in organic chemistry. It is fruitful to note that if we cannot yet consider them cells, some of their properties (particularly their faculty of multiplying in contact with living tissue) definitely foreshadow those of proper organic beings.[1]

Thanks to the discovery of these giant corpuscles the foreseen

[1] Since the viruses have now become *visible* under the powerful magnification of the electron microscope in the form of fine rods asymmetrically active at their two extremities, the opinion has gained ground that we should include them among bacteria rather than among 'molecules'. But then, surely, the study of enzymes and other complex chemical substances is beginning to reveal that molecules have a *form* and even a great variety of forms.

existence of *intermediate states* between the microscopic living world and the ultra-microscopic 'inanimate' one has now passed into the field of direct experimentation.

So from now on we are justified not only by our intellectual need of continuity but by positive indications when we state that, in accordance with our theoretical anticipation of the reality of a pre-life, some natural function really does link the mega-molecular to the micro-organic both in the sequence of their appearance and in their present existence.

And this preliminary finding takes us another step towards a better understanding of the preparations for, and hence the origins of, life.

### B. *A Forgotten Era*

I am not enough of a mathematician to be able to judge either the well-foundedness or the limits of relativity in physics. But, as a naturalist, I am obliged to recognise that the assumption of a dimensional *milieu* in which space and time are organically combined is the only way we have found to explain the distribution around us of animate and inanimate substances. Indeed the further we advance in our knowledge of the natural history of the world, the more clearly we realise that the distribution of objects and forms at any given moment can only be explained by a process whose duration in time varies directly with the spatial (or morphological) dispersion of the objects in question. Every distance in space, every morphological deviation, presupposes and expresses a duration.

Let us take the very simple case of existing vertebrates. In the time of Linnaeus the classification of these animals had advanced sufficiently for them to be arranged in a definite structure of orders, families, genera etc. Yet the naturalists of the day were unable to provide any scientific explanation of this system. We know now that the system of Linnaeus merely represents a present-day cross-section of a diverging bundle of *phyla*[1] emerging

[1] [Throughout this work, the author uses the word *phylum* in its looser sense for a zoological branch regardless of dimension.]

one after the other through the centuries.[1] Accordingly the zoological separation of living creatures into different types reveals and measures in each case a difference in age. In the constellation of species, everything which exists and the place which it occupies implies a certain past, a certain genesis. In particular every time the zoologist meets a more primitive type than those he is familiar with (take the amphioxus, for example) the result is not merely to extend by one more unit the range of animal forms: no, a discovery of that sort *ipso facto* implies another stage, verticil, or ring on the tree-trunk of evolution. For the amphioxus we can only find a place in the present animal kingdom by supposing a whole ' proto-vertebrate ' stage of life in the past, coming somewhere beneath the fishes.

*In the biologist's space-time, the introduction of a new morphological end-form or stage needs immediately to be translated by a correlative prolongation of the axis of duration.*

Keeping this principle in mind, let us return to these astonishing giant molecules detected by recent science.

It is possible, though unlikely, that these enormous particles form in nature today no more than an exceptional and relatively restricted group. But however rare they may be, and however modified by secondary association with the living tissue they batten on parasitically, we have no right whatever to treat them as monstrosities or aberrant forms. On the contrary, everything points to their being representative forms, even if only as a surviving residue of some particular stage in the construction of terrestrial matter.

Thus, between our cellular zone and our molecular zone, hitherto supposed adjacent, another, the mega-molecular zone, has now insinuated itself. And at the same time, because of the close relation we have established between space and duration, an additional period must accordingly be inserted at some point far behind us in the history of the earth. Another circle on the trunk of the tree means another interval of time in the life of the

[1] See what I have to say on this subject in the next chapter, section 3, *The Tree of Life.*

universe. The discovery of viruses and other similar elements not only adds another and important term to our series of states and forms of matter; it obliges us to interpolate a hitherto forgotten era (an era of sub-life) in the series of ages that measure the past of our planet.

Accordingly, working down from incipient life, we find once again in a clearly defined terminal form that phase and that aspect of the early earth which we were led to suppose earlier on when we were climbing the ladder of multiple elements.

Naturally we are not yet in a position to say anything definite concerning the length of time required for the establishment of the mega-molecular world. But though we cannot put it into figures, there are nevertheless some considerations to help us to form an idea of its order of magnitude. Here are three reasons among others for believing the process to have been one of the utmost slowness.

In the first place, its appearance and development must have been narrowly dependent on the transformation of the general conditions, chemical and thermal, prevailing on the surface of the planet. In contrast to life, which seems to have spread with an inherent speed in practically stable material surroundings, the mega-molecules must have developed according to the earth's *sidereal* rhythm, i.e. incredibly slowly.

Secondly, the transformation, once begun, must have extended to a mass of matter sufficiently important and sufficiently large to constitute a zone or envelope of telluric dimensions before it could form the necessary basis for the emergence of life. That, too, must have taken a very long time.

Thirdly, mega-molecules seem to show traces of a long history. How could we possibly imagine them forming suddenly, like the simpler corpuscles, and remaining so once and for all? Their complication and their instability, rather like those of life, both suggest a long process of gradual accretions over a series of generations.

For these three reasons, we may now hazard the guess that the duration required for the formation of proteins on the surface of the earth was as long as, perhaps longer than, the whole of geological time from the Cambrian period to the present day.

And so the abyss of the past is deepened by yet another level or layer ; and though our incurable intellectual weakness encourages us to compress it into an ever thinner slice of duration, scientific analysis is constantly forcing us to enlarge it.

This gives us the sort of basis we need for the views which follow.

Without a long period for maturing no profound change can take place in nature. On the other hand, granted such a period, it is inevitable that something *quite new* should be produced. A terrestrial era of the mega-molecule is not merely a supplementary period added to our schedule of durations. For something much more than that is involved, namely the requirement of a critical point which concludes and closes it. Which is exactly what we need to justify the idea that an evolutionary break of the first order must have taken place with the appearance of the first cells.

But in what way can we envisage the nature of this break ?

### C. *The Cellular Revolution*

*a. External Revolution.* From an external point of view, which is the ordinary biological one, the essential originality of the cell seems to have been the discovery of a new method of agglomerating a larger amount of matter in a single unit. This discovery was doubtless prepared over a long period by the tentative gropings in the course of which the mega-molecules gradually emerged ; but for all that it was sufficiently sudden and revolutionary to have immediately enjoyed prodigious success in the natural world.

We are still a long way from being able to define the basic principle of cellular organisation, though it is probably clarity itself. We have, however, learnt enough to be able to estimate the extraordinary complexity of its structure and the no less extraordinary fixity of its fundamental type.

First the *complexity*. Chemistry teaches us that the cellular edifice is based on albuminoids, nitrogenous organic substances (amino acids) of enormous molecular weight (up to 10,000 and over). In combination with fats, water, phosphorus, and all sorts of mineral salts (potassium, sodium, magnesium, and various metallic compounds) these albuminoids constitute a 'protoplasm', a sponge made up of innumerable particles in which come appreciably into play the forces of viscosity, osmosis, and catalysis which characterise matter when molecular groupings have reached an advanced stage. And that is not all. In the centre of this agglomeration a nucleus containing 'chromosomes' may generally be seen against the background of the surrounding 'cytoplasm', perhaps itself composed of fine rods or filaments ('mitochondria'). With the increased powers of the microscope and advances in the use of stains, new structural elements continue to appear in the complex (whether in height or depth). We find a triumph of multiplicity organically contained within a minimum of space.

Next the fixity. As we have already pointed out, indefinite as are the possible modulations of the fundamental theme, inexhaustible as are the various forms it assumes in nature, the cell remains in all cases essentially true to itself. Looking at it, we hesitate to compare it to anything either in the world of the 'animates' or that of the 'inanimates'. Yet cells still seem to resemble one another more as molecules do than as animals do. We are right to look on them as the first of living forms. But are we not equally entitled to view them as the representatives of *another state* of matter, something as original in its way as the electronic, the atomic, the crystalline, or the polymerous ? As a new type of material for a new stage of the universe ?

In this cell (at the same time so single, so uniform and so complex) what we have is really the stuff of the universe reappearing once again with all its characteristics—only this time it has reached a higher rung of complexity and thus, by the same stroke (if our hypothesis be well founded), advanced still further in *interiority*, i.e. in consciousness.

*b. Internal Revolution.* It is generally accepted that we must assume psychic life to ' begin ' in the world with the first appearance of organised life, in other words, of the cell. I am thus at one with current views and ways of stating them when I assume a decisive step in the progress of consciousness on earth to have taken place at this particular stage of evolution.

But since I have admitted a much earlier origin (a primordial one in fact) to the first lineaments of immanence within matter, it is incumbent on me to explain in what specific way the internal (' radial ') energy is modified to correspond with the external (' tangential ') constitution of the cellular unit. If we have already endowed the long chain of atoms, then molecules, then mega-molecules, with the obscure and remote sources of a rudimentary free activity, it is not by a totally new beginning but by a *metamorphosis* that the cellular revolution should express itself psychically. But how ? How are we to envisage the change-over (how are we even to find room for a change-over) from the pre-consciousness inherent in pre-life to the consciousness, however elementary, of the first true living creature ? Are there several ways for a creature to have a *within* ?

It is not easy, I must confess, to be clear on this point. Later on, in the case of thought, a psychical definition of the ' human critical point ' will emerge almost at once, because the threshold of reflection bears in itself something definitive and also because we have only to consult our own deeper selves to measure it. If, on the other hand, we wish to compare the cell with its predecessors, introspection can only help us through repeated and remote analogies. What do we know of the ' souls ' of animals, even of those nearest to ourselves ? At such distances downward and backward we must resign ourselves to being vague in our speculations.

At grips with this obscurity and marginal approximation, we are nevertheless able to make at least three possible observations—which are enough to fix in a useful and coherent way the position of the *cellular awakening* in the series of psychical transformations preparing the advent on earth of the phenomenon of man.

Even if we accept that a sort of rudimentary consciousness precedes the emergence of life, especially if we accept it, such an awakening or jump (i) *could*, or, better, (ii) was *bound to*, happen, and hence (iii) we have a partial explanation for one of the most extraordinary renewals which the face of the earth has undergone historically.

In the first place it is quite conceivable that an essential change-over between two states or forms of consciousness, even on the lower levels, can happen. To return to and change round in its very terms the doubt formulated above, I would say there were a good many ways for a being to have a ' *within* '. A *closed* surface, irregular at first, may become *centred*. A circle can augment its order of symmetry and become a sphere. Either by arrangement of the parts or by the acquisition of another dimension, the degree of ' interiority ' of a cosmic element can undoubtedly vary to the point at which it rises suddenly on to another level.

Now that precisely such a psychic mutation must have accompanied the discovery of cellular combination follows directly from the law accepted above as regulating the mutual relations of the *within* and the *without* of things. The increase of the synthetic state of matter involves, we said, a corresponding increase of consciousness for the *milieu* synthesised. To which we should now add: the *critical* change in the intimate arrangement of the elements induces *ipso facto* a change of *nature* in the state of consciousness of the particles of the universe.

And now, in the light of these principles, let us look once again at the astounding spectacle displayed by the definitive budding of life on the surface of the early earth ; at the thrust forward in spontaneity ; at the luxuriant unleashing of fanciful creations ; at the unbridled expansion and the leap into the improbable. Surely the explosion of internal energy consequent upon and proportioned to a fundamental super-organisation of matter is precisely the event which our theory could have led us to expect.

Such an external realisation of an essentially new type of corpuscular grouping, allowing the more supple and better centred organisation of an unlimited number of substances at all and, simultaneously, the internal onset of a new type of conscious activity and determination : this double and radical metamorphosis allows us reasonably to define, in regard to what is specifically original in it, the critical passage from the molecule to the cell—the transit to life.

Before considering the subsequent evolutionary consequences of this transit, we must look a little closer into the conditions of its historical realisation—firstly in space, and secondly in time.

That is the object of the two sections which follow.

## 2. THE INITIAL MANIFESTATIONS OF LIFE

Because the apparition of the cell was an event which took place on the frontiers of the infinitesimal, and because the elements involved were delicate in the extreme, now absorbed in sediments transformed long ago, there is no chance, as I have said already, of our ever finding traces of it. Thus at the outset we come up against that fundamental condition to which experience is subject, by virtue of which the beginnings of all things tend to be materially out of our grasp. This is a law running right through history which we shall later be calling the ' automatic suppression of evolutionary peduncles '.

Fortunately there are a number of different ways in which our minds can reach reality. What escapes the intuition of our senses we can encircle and define approximately by a series of indirect attacks. Let us follow this more roundabout method, the only one at our disposal when we try to picture new-born life. We can do so by stages in the following manner.

### A. *The* Milieu

We must start by going back perhaps a thousand million years and wipe out the greater part of those material superstructures which form the features of the earth's surface today. Geologists are far from being agreed upon what our planet looked like at that distant period. I am inclined myself to picture it as enveloped in a shoreless ocean (of which the Pacific is perhaps a vestige) through which, at a few isolated points, protuberances of future continents had begun to emerge by volcanic eruption. Those waters were doubtless warmer than our seas today and also more fraught with free valencies that succeeding ages were gradually to absorb and stabilise. It was in such a liquid, heavy and active —at all events it was inevitably in a liquid environment— that the first cells must have formed. Let us try to distinguish them.

At this distance of time their form can only be vaguely surmised. By analogy with what we must assume to be their least altered traces today, the best we can do is to imagine this primordial generation in terms of granules of protoplasm, with or without an individually differentiated nucleus. But if the outline and individual structure remain inscrutable, certain characteristics of another order stand out sharply and lose none of their value because they are quantitative. I am referring to their incredible smallness and—natural consequence—their bewildering number.

### B. *Smallness and Number*

Having reached this point we must force ourselves to make one of those 'efforts to see' that I mentioned in my Foreword. We can look at the night sky year in year out without ever once making a *real* effort to apprehend the distances and thus the vast size of the sidereal masses. Similarly our eyes may be familiar with the field of vision of a microscope without our ever 'realising' the disconcerting dimensional hiatus which separates the world of mankind from that of a drop of water. We can speak with accuracy about creatures measurable in hundredths of a millimetre, but have we ever attempted to transplant them mentally seeing them on their own scale in our framework ? Yet this effort at perspective is indispensable if we wish to probe the secrets or even the 'space' of nascent life which can of course be nothing else than a *granular life*.

That the first cells were infinitesimal there can be no doubt. That is determined by their originating out of mega-molecules. It is also established visually when we examine the simplest forms of life that we can find still today in the world. When we finally lose sight of bacteria they are no more than one five-thousandth of a millimetre long.

And there seems positively to be in the universe a natural relationship between size and number. Either because they are faced with a relatively greater space or else to compensate for their reduced effective radius of individual action, the smaller creatures are the more they swarm. Measurable only in terms of microns, the first cells must have been numbered by the myriad. Hence as we get as near as we can to the threshold of life, it manifests itself to us *simultaneously* as *microscopic and innumerable*.

There is nothing in this which should surprise us. Surely it is natural that life, as it just emerges from matter, should be 'dripping with molecularity'.

What we need now is to understand how the organic world works and what is its future. On the bottom rung of that ladder we find number, an immense number. How are we to picture the historical modalities and the evolutive structure of this native multiplicity ?

### C. *The Origin of Number*

From our remote standpoint it may be said that life no sooner started than it swarmed.

To explain and make clear the nature of this multiplicity from the very beginning of animate evolution, two lines of thought suggest themselves.

First of all we can assume that, though they only occurred in the first instance at a single point or a small number of points, the first cells multiplied almost instantaneously—as crystallisation spreads in a super-saturated solution. For surely the early earth was in a state of biological super-tension.

Or, on the other hand, we can equally well suppose that the passage from mega-molecule to cell took place simultaneously at a great many points, the requisite conditions of instability being widespread. Just as, in the case of mankind, great discoveries are often simultaneous.

Was the origin of cells ' monophyletic ' or ' polyphyletic ' ? Was this advance in the first instance simple and narrow but broadening outwards with extreme rapidity, or on the contrary relatively broad and complex from the first and subsequently spreading more slowly ? Which is the most suitable way of imagining the beginnings of the bundle of living beings ?

All through the story of the organisms, at the start of each zoological group, we meet the same problem—single thread or multiple strand ? And just because the beginnings are always beyond the reach of direct vision, we constantly face the same difficulty of choosing between two hypotheses which are almost equally plausible. This hesitation worries and irritates us.

But do we really need to choose—here at any rate ? However slender we may suppose it, the initial peduncle of terrestrial life must have contained an appreciable number of fibres rooted in the enormity of the molecular world. Conversely, however broad we imagine its section, it must, like all nascent physical realities, have enjoyed an exceptional aptitude to branch out into new forms. Fundamentally the two perspectives differ only in the relative importance attributed to one or other of the two factors (initial complexity and ' expansiveness ') present in both cases. Both, moreover, imply a *close relationship* of an evolutive kind between the first living objects on the early earth. So, ignoring their secondary conflicts, let us concentrate on the essential fact on which they both cast light. This, in my opinion, may be expressed as follows :

From whatever angle we look at it, the *nascent* cellular world shows itself to be already infinitely complex. Either on account of its multiple origin, or because of its rapid variegation from a very few points of emergence, or again, we must add, because of regional differences (climatic or chemical) in the earth's watery envelope, we are led to envisage life on the protocellular level as an enormous bundle of polymorphous fibres. Already and even at these depths the phenomenon of life cannot be really understood except as an organic problem of masses in movement.

An organic problem of masses or multitudes and not a simple statistical problem of large numbers : what does that difference imply ?

### D. *Inter-relationship and Shape*

Once more, but now on the collective scale, we are faced with the frontier between the physical and the biological worlds. As long as we were dealing with churning atoms or molecules we could be content with the numerical laws of probability when working out the behaviour of matter. But from the moment when the monad acquires the dimensions and superior spontaneity of a cell, and tends to be individualised at the heart of a pleiad, a more complicated pattern appears in the stuff of the universe. On two counts at least it would be inadequate and false to imagine life, even taken in its granular stage, as a fortuitous and amorphous proliferation.

Firstly the initial mass of the cells must from the start have been inwardly subjected to a sort of inter-dependence which went beyond a mere mechanical adjustment, and was already a beginning of ' symbiosis ' or life-in-common.

However tenuous it was, the first veil of organised matter spread over the earth could neither have established nor maintained itself without some network of influences and exchanges which made it a biologically *cohesive* whole. From its origin, the cellular nebula necessarily represented, despite its internal multiplicity, a sort of diffuse super-organism. Not merely a *foam of lives* but, to a certain extent, itself a *living film*. A simple reappearance, after all, in more advanced form and on a higher level of those much older conditions which we have already seen presiding over the birth and equilibrium of the first polymerised, substances on the surface of the early earth. A simple prelude too, to the much more advanced evolutionary solidarity, so marked in the higher forms of life, whose existence obliges us increasingly to admit the strictly organic nature of the links which unite them in a single whole at the heart of the *biosphere*.

Secondly (and this is more surprising) the innumerable elements composing at the outset the living film of the earth do not seem to have been taken or collected exhaustively and haphazard. Their admission into this primordial envelope gives rather the impression of having been mysteriously guided by a previous selection or dichotomy. Biologists have noted that, according to the chemical group to which they belong, the molecules incorporated into living matter are all asymmetrical in the same way, that is to say if a pencil of polarised light is passed through them they all turn the plane of the beam *in the same direction*—either they are all right-rotating or all left-rotating according to the group taken. More remarkable still, all living creatures, from the humblest bacteria to man, contain exactly the same complicated types of vitamins and enzymes, notwithstanding the great range of chemical forms possible; just as the higher mammals are all ' tritubercular ' and walking vertebrates all four-footed. Surely such similarity of living substance in dispositions *which do not seem necessary* suggests an early choice or sorting. This chemical uniformity of protoplasm at accidental points has been taken as proof that all existing organisms descend from a single ancestral group (the case of the crystal falling in the super-saturated solution). Without going as far as that, we may say that all it establishes is a certain initial cleavage (between right-rotating and left-rotating examples, for instance, whichever it may be) in the enormous mass of carbon matter at the threshold of life (instance of the discovery in *n* points at once). In any event, it is not important. The interesting thing is that on either assumption the living world assumes the same curious appearance of a totality re-formed from a *partial* group : whatever may have been the complexity of its original impetus, it exhausts *only a part of what might have been.* Taken as a whole, the biosphere would thus represent only a simple *branch* within and above other less progressive or less fortunate proliferations of pre-life. And surely this amounts to saying that, considered globally, the appearance of the first cells gives rise to the same problems as do the origins of each of those later stems we call ' phyla '. The universe had *already begun to ramify* and it doubtless goes on ramifying indefinitely, *even below* the tree of life.

Seen from afar, elementary life looks like a variegated multitude of microscopic elements, a multitude great enough to envelop the earth, yet at the same time sufficiently interrelated and selected to form a structural whole of genetic solidarity.

These remarks, let it be said again, are only valid for the general features and characters taken as a whole. That is what should have been expected and we must be resigned to it. Following all the dimensions of the universe one same law of perspective inevitably blurs, in the field of our vision, the abysses of the past and the distant backgrounds of space : what is very far and very small loses its outline. For us to probe further into the phenomena accompanying its origin, it would be necessary for life—somewhere or other on the earth—to be still generating today under our eyes.

That chance—and here is my last point under this heading—is precisely the one we are not given.[1]

## 3. THE SEASON OF LIFE

It would be quite conceivable *a priori* that the mysterious transformation of mega-molecules into cells, accomplished millions of years ago, might still, unnoticed, be going on around us at the extreme limits of the microscopic and the infinitesimal. There are many forces in nature that we have supposed exhausted only to find, on closer analysis, that they are still flourishing. The earth's

[1] Unless of course (and who can tell?) chemists succeed in reproducing the phenomenon in the laboratory.

crust has not yet stopped heaving and plunging under our feet. Mountain ranges are still being thrust up on the horizon. Granites are still growing under the continental masses. Nor has the organic world ceased to produce new buds at the tips of its countless branches. If movement can be concealed by extreme slowness, why should not extreme smallness have the same effect? Indeed there is nothing inherently impossible about the continued birth today of living substance on an infinitesimal scale.

In fact, however, nothing indicates this to be the case. On the contrary, everything points the other way.

We all know of the famous controversy of nearly a hundred years ago between the partisans and the adversaries of 'spontaneous generation'.. It would appear that too much was made at the time of the results of the battle, as though Pouchet's defeat closed the door on any scientific hope of giving an evolutionary explanation to the first origins of life. But today we are all agreed on one point. From the fact that, in the laboratory, life never appears in a medium from which all germs have previously been eliminated, it would be a mistake to deduce (in the face of all manner of general evidence) that the phenomenon may not have happened under other conditions in other ages. Pasteur's experiments could not and cannot now in any way disprove the birth of cells on our planet in the past. But their success, proved over and over again by the universal adoption of methods of sterilisation, seems to have really established one thing: that within the field and limits of what we can investigate, protoplasm is *no longer formed directly* from the inorganic substances of the earth.[1]

This obliges us at the outset to revise certain over-absolute ideas we may have harboured concerning the use and value in our sciences of explanations '*in terms of present causes*'.

[1] Against Pasteur's experiments it may be objected that sterilisation is so brutal as to be capable of destroying not only the living germs, whose elimination is desired, but also those 'pre-living' germs from which alone life might emerge. However that may be, the most convincing proof to me that life was produced once and once only on earth is furnished by the profound structural unity of the tree of life (see below).

A moment ago I reminded the reader that many terrestrial transformations which we could have sworn had stopped, and stopped ages ago, are still going on in the world around us. Under the influence of this unexpected observation which pampers our natural preference for palpable and manageable forms of experience, our minds are inclined to slide gently into the belief that there never was in the past or will be in the future anything new under the sun. And it would only be one step farther to limit full and real knowledge to the events of the present. Fundamentally, is not everything, apart from the present, mere 'conjecture'?

We must at all costs resist this instinctive limitation of the rights and scope of science.

No. The world would not fully satisfy the conditions imposed by actuality—it would not be the great world of mechanics and biology—if we were lost in it like ephemeral insects which are unaware of all save their brief season. So vast are the dimensions of the universe disclosed by the present that, for this reason alone, all sorts of things must have happened in it before man was there to witness them. Long before the awakening of thought on earth, manifestations of cosmic energy must have been produced which have no parallel today. Thus, besides the group of phenomena subject to direct observation, there is for science a particular class of facts to be considered—specifically the most important because the rarest and most significant—those which depend neither on direct observation nor experiment, but can only be brought to light by a very authentic branch of 'physics', the *discovery of the past*. And, to judge by our repeated failures to find its equivalent around us or to reproduce it, the first apparition of living bodies is clearly one of the most sensational of these events.

With that, let us advance a step. There are two possible ways in which something can fail to coincide, in time, with our power of seeing. One is for it to happen at such distant intervals that the whole of our existence can run its course between two successive occurrences. The other, by which we miss it still more inevitably, is for it to have happened once and never be repeated. In other words, either a recurrent phenomenon of very infrequent periodicity (such as we meet so often in astronomy) or one strictly unique (as with Socrates or Augustus in human history). In which of these two 'inexperimental' or rather 'praeter-experimental' categories do we find it most suitable, in the light of Pasteur's discoveries, to put the birth of life, the initial formation of cells from matter ?

There is no lack of facts to support the idea that organised matter might germinate *periodically* on the earth. Later on, when I come to outline the 'tree of life', I shall be calling attention to the coexistence in the living world of certain large aggregates (protozoa, plants, hydrozoa, insects, vertebrates) whose lack of basic relationship might be fairly satisfactorily explained in terms of heterogenous origins. Something like those successive intrusions going back to different ages originating from the same magma, whose interlacing veins form the eruptive complex of a single identical mountain . . . the hypothesis of independent vital pulsations would conveniently account for the morphological diversity of the principal sub-kingdoms recognised by systematic biology. Moreover, there is no difficulty on the chronological side. In any case the length of time separating the historical origins of two successive sub-kingdoms is much greater than the age of mankind. So it is not astonishing that we should live in the illusion that nothing happens any more. Matter seems dead. But could not the next pulsation be slowly preparing around us ?

I feel bound to point out and even, to a certain extent, to defend the conception of a spasmodic genesis of life. Yet I cannot actually adopt it. For there is one decisive objection against the idea of a number of different, successive, vital thrusts on the earth's surface—namely the fundamental similarity of all organic beings.

We have already called attention in this chapter to the curious fact that all molecules of living substances are asymmetrical *in the same way*, and contain precisely the same vitamins. Now, the more complex organisms become, the more evident becomes their inherent kinship. It manifests itself in the absolute and universal uniformity of the basic cellular pattern, and it manifests itself, particularly in animals, in the identical solutions found for various problems of perception, nutrition and reproduction—everywhere we find vascular and nervous systems, everywhere some form of blood, everywhere gonads and everywhere eyes. It continues in the similarity of the methods employed by units for collecting together in higher organisms and becoming 'socialised', and finally it shines clearly in the general laws of development ('ontogenesis' and 'phylogenesis') which give to the living world, considered as a whole, the coherence of a single upthrust.

Though one or the other of these many analogies might be explained by the adjustment of one and the same 'pre-living magma' under identical terrestrial conditions, it would nevertheless seem impossible to regard their unified complex as the result of a simple parallelism or a simple 'convergence'. Even if there were only one solution to the main physical and physiological problem of life on earth, that general solution would necessarily leave undecided a host of accidental and particular questions, and it does not seem thinkable that they would have been decided *twice in the same way*. And it is precisely in these ancillary modalities that living creatures resemble each other, even those belonging to very different groups. Accordingly the contrasts presented today by zoological phyla lose much of their importance (are they not simply effects of perspective combined with a progressive isolation of existing phyla ?), and naturalists are becoming more and more convinced that the genesis of life on earth belongs to the category of absolutely *unique* events that, once happened, are never repeated. This is a much more credible hypothesis than would appear at first sight, if we succeed in forming a tenable idea of what is hidden in the history of our planet.

It is fashionable nowadays in geology and geophysics to attach a preponderant importance to periodical phenomena. Seas advance and recede ; continental platforms rise and sink ; mountains are lifted and levelled ; glaciations advance and retire;

radio-active warmth accumulates in the depths then overflows on the surface. We hear of nothing save this majestic 'ebb and flow' in treatises dealing with the vicissitudes of the earth.

This predilection for what is rhythmic in events goes hand in hand with a preference for the 'actual' in causes, and both alike are explained by precise rational needs. Whatever repeats itself is, at all events potentially, observable, and can be made subject to a law. It provides a scale on which we can measure time. I am the first to acknowledge the scientific quality of these advantages, yet I cannot help thinking that an exclusive analysis of the oscillations recorded by the earth's crust or the movements of life would omit from the inquiry what is the principal aim of geology.

For the earth is after all something more than a sort of huge breathing body. Admittedly it rises and falls, but more important is the fact that it must have begun at a certain moment; that it is passing through a consecutive series of moving equilibria; and that in all probability it is tending towards some final state. It has a birth, a development, and presumably a death ahead. Thus all around us, deeper than any pulsation that could be expressed in geological eras, we must suppose there to be a total process which is not of a periodic character defining the *total* evolution of the planet; something more complicated chemically and deeper within matter than the 'cooling' of which we used to hear so much; yet something both continuous and irreversible. An ever-ascending curve, the points of transformation of which are never repeated; a constantly rising tide below the rhythmic tides of the ages—it is on this essential curve, it is in relation to this advancing level of the waters, that the phenomenon of life, as I see things, must be situated.

If life, one day, was able to 'isolate' itself in the primitive ocean, it was no doubt because the complexity of the earth's elements and their distribution had reached the general privileged condition which permitted and favoured the building of protoplasms (which is what we mean by the earth being 'young').

And if thereafter life has never again been formed directly

from the elements of the lithosphere or hydrosphere, this is apparently because the very emergence of a biosphere so disturbed, impoverished and relaxed the primordial chemism of our fragment of the universe that the phenomenon can never be repeated (unless perhaps artificially).

From this point of view—and it seems to me the right one—the ' cellular revolution ' would now be seen as a critical singular point, an unparalleled moment on the curve of telluric evolution, the point of *germination*. Protoplasm was formed once and once only on earth, just as nuclei and electrons were formed once and once only in the cosmos.

This hypothesis has the advantage of providing a reason for the deep organic likeness which stamps all living creatures from bacteria to mankind. At the same time it explains why we never at any point find the formation of the least living thing which is not there as the result of generation. And that was the problem.

But this hypothesis has two other notable consequences for science.

Firstly, by separating the phenomenon of life from the numerous other periodical and secondary events on earth, and by making it one of the principal landmarks (or parameters) of the sidereal evolution of the globe, it rectifies our sense of proportion and of values and hence renews our perspective of the world.

Secondly, by the very fact of showing that the origin of organised bodies is linked with a chemical transformation unprecedented and unrepeated in the history of the world, the hypothesis inclines us to think of the energy contained in the living layer of our planet as developing from and within a sort of closed ' quantum ', defined by the amplitude of this primordial emission.

Life was born and propagates itself on the earth as a solitary pulsation.

It is the propagation of that unique wave that we must now follow, right up to man and if possible beyond him.

CHAPTER TWO

# THE EXPANSION OF LIFE

WHEN A physicist wants to study the development of a wave, he begins by calculating the pulsation of a single particle. Then he reduces the vibrating medium to its main characteristics and directions of elasticity, and generalises the results found in the instance of the element. He thus obtains an overall picture as close as possible to the movement of the whole he is trying to determine.

When he faces the task of describing the ascent of life, the biologist is obliged to follow a similar method in his own special way. It is impossible to reduce this enormous and complex phenomenon to order without first analysing the processes discovered by life for its advance in each of its elements taken in isolation. It is equally impossible to distinguish the general behaviour adopted by the total multitude of individual progressions without choosing the most expressive and luminous features of their resultant effect.

In the pages that follow I intend to develop a simplified but structural representation of life evolving on earth ; a vision so homogeneous and coherent that its truth is irresistible. I provide no minor details and no arguments, but only a perspective that the reader may see and accept—or not see.

The gist of what I mean comes under these three headings :

1. The elemental movements of life,
2. The spontaneous ramification of the living mass,
3. The tree of life.

All this will first be studied at the surface and from *without*. We shall only start probing into the *within* of things in the subsequent chapter.

## 1. THE ELEMENTAL MOVEMENTS OF LIFE

### A. *Reproduction*

At the base of the entire process whereby the envelope of the biosphere spreads its web over the face of the earth stands the mechanism of reproduction which is typical of life. Sooner or later each cell divides (by mitotic or amitotic division) and gives birth to another cell similar to itself. First, a single centre ; then two. Everything in the subsequent development of life stems from this potent primordial phenomenon.

In itself, cell division seems to be due to the simple need of the living particle to find a remedy for its molecular fragility and for the structural difficulties involved in continued growth. The process is one of rejuvenation and shedding. The more limited groups of atoms, the micro-molecules, have an almost indefinite longevity, and with it an equivalent rigidity. The cell, continually in the toils of assimilation, must split in two to continue to exist. At first sight reproduction appears as a simple process thought up by nature to ensure the permanence of the unstable in the case of these vast molecular edifices.

But, as always happens in the world, what was at first a happy accident or means of survival, is promptly transformed and used as an instrument of progress and conquest. Life at first seems to have reproduced itself only in self-defence ; but this was a mere prelude to its vast conquests.

### B. *Multiplication*

For, once introduced into the stuff of the universe, the principle of the duplication of living particles knows no limits other than those of the quantity of matter provided. It has been calculated that, in a few generations, a single infusorian could by simple division of itself and its descendants cover the whole surface of the earth. Every volume, however great, succumbs to the effects of geometrical progression, and this is not a pure extrapolation of the mind. In its ability to double itself and to go on doubling itself without let or hindrance, life possesses a force of expansion as invincible as that of a body that dilates or vaporises. But whereas in the case of so-called inert matter the increase in volume soon reaches a point of equilibrium, no such limit appears to be set to the expansion of living substance. The more the phenomenon of cellular division spreads, the more it gains in virulence. Once fission has started, nothing from within can arrest its devouring and creative conflagration, because it is spontaneous. Nor is there any external influence powerful enough to check the process.

### C. *Renovation*

Yet this is only the first result and only the quantitative side of the process. Reproduction doubles the mother cell. Thus, by a mechanism which is the inverse of chemical disintegration, *it multiplies without crumbling*. At the same time, however, it transforms what was only intended to be prolonged. Closed in on itself, the living element reaches more or less quickly a state of immobility. It becomes stuck and coagulated in its evolution. Then by the act of reproduction it regains the faculty for inner re-adjustment and consequently takes on a new appearance and direction. The process is one of pluralisation in form as well as in number. The elemental ripple of life that emerges from each individual unit does not spread outwards in a monotonous circle formed of individual units exactly like itself. It is diffracted and becomes iridescent, with an indefinite scale of variegated tonalities. The living unit is a centre of irresistible multiplication, and *ipso facto* an equally irresistible focus of diversification.

### D. *Conjugation*

And then, so it seems, so as to enlarge the breach thus made by its first inroads in the ramparts of the unorganised world, life discovered the wonderful process of conjugation. It would take a whole book to describe and extol the growth and sublimation of sexual dualism in the course of evolution from the cell to man. At the early stages that we are now considering, the phenomenon appears in the main as a means of accelerating and intensifying the double effect (multiplication and diversification) obtained by a sexual reproduction such as is still prevalent in many of the lower organisms and even with the individual cells of our own bodies. By the first conjugation of two elements, however little they may as yet have been differentiated into male and female, the door was thrown open to those modes of generation whereby a single individual can pulverise itself into a myriad of germs. Simultaneously we find coming into play the endless permutations and combinations of 'characters' so dear to modern geneticists. Instead of simply radiating from each centre in process of division, the rays of life now anastomose—exchanging and varying their respective riches. We no more dream of being astonished at this prodigious invention than at the discoveries of fire, bread or writing. Yet what chances and what fumblings—and what endless ages therefore—were necessary before this fundamental discovery from which we have sprung was matured. And how much longer still before it found its complement and natural fulfilment in the no less revolutionary innovation of 'association'.

### E. *Association*

In first analysis—and supposing we ignore deeper factors for the moment—the grouping of living particles into complex organisms is an almost inevitable consequence of their multiplication. Cells tend to congregate because they press against each other or are even born in clusters. But the purely mechanical necessity or opportunity to get together engendered in the long run a definite method of biological improvement.

We still seem to be able to see all the stages of this *still unfinished* march of nature towards the unification or synthesis of the ever-increasing products of living reproduction. At the bottom we find the simple aggregate, as in bacteria and the lower fungi. One stage higher comes the colony of attached cells, not yet centralised, though distinct specialisation has begun, as with the higher vegetable forms and the bryozoa. Higher still is the metazoan cell of cells, in which by a prodigious critical transformation an autonomous centre is established (as though by excessive shrinking) over the organised group of living particles. And still farther on, to round off the list, at the present limit of our experience and of life's experiments, comes *society*—that mysterious association of free metazoans in which (with varying success) the formation of hyper-complex units by 'mega-synthesis' seems to be being attempted.

The last part of this book will be particularly devoted to this last and highest form of aggregation, in which the self-organising effort of matter culminates perhaps in society as capable of reflection. Here we must confine ourselves to pointing out that association, considered at all its levels, is not a sporadic or accidental appearance in the animal kingdom. On the contrary, it represents one of the most universal and constant expedients (and thus one of the most significant) used by life in its expansion. Two of its advantages are immediately obvious. Thanks to it, living substance is able to build itself up in sufficient bulk to escape innumerable external obstacles (capillary attraction, osmotic pressure, chemical variation of the medium, etc.) which paralyse the microscopic organisms. In biology, as in navigation, a certain size is physically necessary for certain movements. Thanks to it again, the organism (here too because of its increased volume) is able to find room inside itself to lodge the countless mechanisms *added successively* in the course of its differentiation.

### F. *Controlled Additivity*

Reproduction, conjugation, association . . . No matter how far they are extended, these various activities of the cell in themselves only lead to a surface deployment of the organisms. If it had been left to their resources alone, life would have spread and varied, but always on the same level. It would have been like an aeroplane which can taxi but not become airborne. It would never have taken off.

It is at this point that the phenomenon of *additivity* intervenes and acts as a vertical component.

There seems to be no lack of examples, in the course of biological evolution, of transformations acting horizontally by pure crossing of characters. One example is the mutation we call Mendelian. But when we look deeper and more generally we see that the rejuvenations made possible by each reproduction achieve something more than mere substitution. They *add*, one to the other, and their sum increases *in a pre-determined direction.* Dispositions are accentuated, organs are adjusted or supplemented. We get diversification, the growing specialisation of factors forming a single genealogical sequence—in other words, the apparition of the *line* as a natural unit distinct from the *individual.* This law of controlled complication, the mature stage of the process in which we get first the micro-molecule then the mega-molecule and finally the first cells, is known to biologists as *orthogenesis.*[1]

Orthogenesis is the dynamic and only complete form of heredity. The word conceals deep and real springs of cosmic extent. We shall find this out little by little, but meanwhile one

[1] On the pretext of its being used in various questionable or restricted senses, or of its having a metaphysical flavour, some biologists would like to suppress the word 'orthogenesis'. But my considered opinion is that the word is essential and indispensable for singling out and affirming the manifest property of living matter to form a system in which 'terms *succeed each other* experimentally, following constantly increasing degrees of centro-complexity'.

point already stands out clearly at the present stage of our inquiry. Thanks to its characteristic additive power, living matter (unlike the matter of the physicists) finds itself 'ballasted' with complications and instability. It falls, or rather rises, towards forms that are more and more improbable.

Without orthogenesis life would only have spread ; with it there is an ascent of life that is invincible.

### *A Corollary : The Ways of Life*

At this point let us pause for a moment. Before we try to see where these various laws regulating the movements of the isolated particle lead us, when extended to the whole of life, let us attempt to distinguish the general lines of behaviour or attitudes which, in accordance with these elementary laws, characterise life in movement at all levels and in all circumstances.

These attitudes or ways of proceeding can be reduced to three: *profusion*, *ingenuity* and (judged from our individual point of view) *indifference*.

*a*. Let us first consider *profusion*, which is born of unlimited multiplication.

Life advances by mass effects, by dint of multitudes flung into action without apparent plan. Milliards of germs and millions of adults jostling, shoving and devouring one another, fight for elbow room and for the best and largest living space. Despite all the waste and ferocity, all the mystery and scandal it involves, there is, as we must be fair and admit, a great deal of biological efficiency in the *struggle for life*. In the course of this implacable contest between masses of living substance in irresistible expansion, the individual unit is undeniably tried to the limits of its strength and resources. 'Survival of the fittest by natural selection' is not a meaningless expression, provided it is not taken to imply either a final ideal or a final explanation.

But it is not the individual unit that seems to count for most in the phenomenon. What we find within the struggle to live is something deeper than a series of duels ; it is a conflict of chances. By reckless self-reproduction life takes its precautions against mishap. It increases its chances of survival and at the same time multiplies its chances of progress.

Once more, this time on the plane of animate particles, we find the fundamental technique of *groping*, the specific and invincible weapon of all expanding multitudes. This groping strangely combines the blind fantasy of large numbers with the precise orientation of a specific target. It would be a mistake to see it as mere chance. Groping is *directed chance*. It means pervading everything so as to try everything, and trying everything so as to find everything. Surely in the last resort it is precisely to develop this procedure (always increasing in size and cost in proportion as it spreads) that nature has had recourse to profusion.

*b*. Next comes *ingenuity*. This is the indispensable condition, or more precisely the constructive facet, of additivity.

To accumulate characters in stable and coherent aggregates, life has to be very clever indeed. Not only has it to invent the machine but, like an engineer, so design it that it occupies the minimum space and is simple and resilient. And this implies and involves, as regards the structure of organisms (particularly the higher ones), a property which must never be forgotten.

*What can be put together can be taken apart.*

At an early stage of their discoveries biologists were surprised and fascinated by the fact that living beings, however perfect (or even more perfect) their spontaneity, were always decomposable into an endless chain of closed mechanisms. From this they thought they could deduce universal materialism. But they overlooked the essential difference between a natural whole and the elements into which it is analysed.

By its very construction, it is true, every organism is always and inevitably reducible into its component parts. But it by no means follows that the sum of the parts is the same as the whole, or that, in the whole, some specifically new value may not emerge. That what is ' free ', even in man, can be broken down into determinisms, is no proof that the world is not based on freedom—as indeed I maintain that it is. It is simply the result of ingenuity—a triumph of ingenuity—on the part of life.

*c.* Lastly, for individual units, comes *indifference*.

How often have artists, poets and even philosophers depicted nature as a blind Fury trampling existence in the dust ?

Profusion is the first trace of this apparent brutality : like Tolstoy's grasshoppers, life passes over a bridge made up of accumulated corpses, and this is a direct effect of multiplication. But in the same ' inhuman ' direction *orthogenesis* and *association* also operate, in their fashion.

By the phenomenon of association, the living particle is wrenched from itself. Caught up in an aggregate greater than itself, it becomes to some extent its slave. It no longer belongs to itself.

And what organic or social incorporation does to extend it in space, its accession to a line of descent achieves no less inexorably in time. By the force of orthogenesis the individual unit becomes part of a chain. From being a centre it is changed into being an intermediary, a link—no longer *existing*, but *transmitting* ; and, as it has been put, life is more real than lives.

On the one hand the individual unit is lost in number, on the other it is torn apart in the collectivity, and in yet a third direction it stretches out in becoming. This dramatic and perpetual opposition between the one born of the many and the many constantly being born of the one runs right through evolution.

As the general movement of life becomes regular, the conflict, despite occasional counter-attacks, tends to resolve itself. Yet it remains painfully noticeable to the end. The antinomy only clears up with the appearance of mind where it attains its paroxysm in *feeling*, and the indifference of the world for its constituents is transformed into an immense solicitude. This is the sphere of the person.

But we have not yet come to that point.

Groping profusion ; constructive ingenuity ; indifference towards whatever is not future and totality ;—these are the three headings under which life rises up by virtue of its elementary mechanisms. There is also a fourth heading which embraces them all—that of *global unity*.

This we have come across already—first in primordial matter, then on the early earth, then in the genesis of the first cells. Here it reappears in a still more emphatic way. Though the proliferations of living matter are vast and manifold, they never lose their *solidarity*. A continuous adjustment co-adapts them from without. A profound equilibrium gives them balance within. Taken in its totality, the living substance spread over the earth—from the very first stages of its evolution—traces the lineaments of one single and gigantic organism.

I repeat this same thing like a refrain on every rung of the ladder that leads to man ; for, if this thing is forgotten, nothing can be understood.

To see life properly we must never lose sight of the unity of the biosphere that lies beyond the plurality and essential rivalry of individual beings. This unity was still diffuse in the early stages —a unity in origin, framework and dispersed impetus rather than in ordered grouping ; yet a unity which, together with life's ascent, was to grow ever sharper in outline, to fold in upon itself, and, finally, to centre itself under our eyes.

## 1. THE RAMIFICATIONS OF THE LIVING MASS

Now let us study, over the whole extent of the living earth, the various movements whose aspect we have analysed in the instance of cells or groups of cells taken in isolation. Seen on such a huge scale one might well expect the multitude to be entangled in utter confusion. Or, inversely, we might expect that their total, in the process of harmonising, should create a continuous wave like the radiating ripple from a stone in a pool. But what actually happens is a third alternative. As we see it under our very eyes today, the 'front' of advancing life is neither chaotic nor continuous. It is an aggregate of fragments at one and the same time divergent and arranged in tiers—classes, orders, families, genera, species. In other words what we see is the whole scale of groups whose variety, order of size and relationships our modern systematic biology tries to express in names.

Considered as a whole, life's advances go hand in hand with segmentation. As life expands, it splits spontaneously into large, natural, hierarchical units. It *ramifies.* And the moment has come to study this ramification, a particular phenomenon as essential to large animate masses as mitotic division was to cells.

A number of different factors contribute to drawing up or accentuating the branches of life. Again, I shall reduce them to three : A. Aggregates of growth, giving birth to 'phyla.' B. Florescence (or disjunctions) of maturity, periodically producing 'verticils'. C. Effects of distance : the elimination (from view) of the 'peduncles'.

### A. *Aggregates of Growth*

Let us return to the living element in the process of reproduction and multiplication. From this element, taken as centre, we have seen different lines radiating orthogenetically, each recognisable by the accentuation of certain characters. By their construction these lines diverge and tend to separate. Yet, so far, we have no reason to suppose that they may not meet with other lines radiating from neighbouring elements, become enmeshed with them and so form an impenetrable network.

By 'aggregate of growth' I mean the new and unexpected fact that a dispersion *of simple type* occurs precisely where the play of chance would have made us most fear a complicated tangle. When poured out on the ground, a sheet of water quickly breaks up into streamlets and then into definite streams. Similarly, under the influence of various causes (such as the native parallelism of elementary orthogenesis, the attraction and mutual adjustment of lines, the selective influence of the environment and so on) the fibres of a living mass in the process of diversification tend to draw together, to bind, following a restricted number of dominant directions. In the beginning this concentration of forms round a few privileged axes is indistinct and indefinite ; it involves a mere increase, in certain sectors, of the number or density of the lines. Then gradually the movement takes shape. True nervures become visible, though without breaking up the limb of the leaf in which they appear. At this stage the fibres may still partially escape from the network which is trying to contain them. From nervure to nervure, they may still touch one another, anastomose, or cross one another. The zoologist would say that the group is still at the racial stage. And at this point there takes place what may be called the final aggregation or final separation (according to the point of view we take). For, having reached a certain degree of mutual cohesion, the lines isolate themselves in a closed sheaf that can no longer be penetrated by neighbouring sheaves. From now on, their association, the ' bundle ', will evolve on its own, autonomously. The species has become individualised. The phylum has been born.

*The phylum*. The living ' bundle '; the line of lines. Many observers still refuse to see or admit the reality of this strand of life in the process of evolution. They do not know how to see, how to make the necessary adjustments in their vision.

The phylum is first of all a collective reality. Therefore, to see it clearly, we need to look from a sufficient height and distance. Examined too closely, it crumbles into unevenness and confusion. We fail to see the wood for the trees.

Secondly, the phylum is polymorphous and elastic. Like a molecule, which ranges through all sizes and degrees of complication, it can be as small as a single species or as vast as a sub-kingdom. There are simple phyla and phyla composed of phyla. Phyletic unity is not so much quantitative as structural ; so we must be ready to recognise it on every scale of dimension.

Lastly, the phylum has a dynamic nature. It only comes properly into view at a certain depth of duration, in other words only in *movement*. When immobilised in time, it loses its features and, as it were, its soul. Its motion is killed by a ' still '.

Considered without these provisos, the phylum might well

be thought to be just one more artificial entity carved for classification purposes out of the continuum of life. But looked at in proper magnification and light, it can be seen to be a perfectly defined structural reality.

What defines the phylum in the first place is its 'initial angle of divergence', that is to say the particular direction in which it groups itself and evolves as it separates off from neighbouring forms.

What defines it in the second place is its 'initial section'. About this point (already touched on when we were considering the first cells, and which will assume outstanding importance in the case of man) we are still very much in the dark. But at least one thing is certain at the outset. Just as it is physically impossible for a drop of water to condense save at a certain volume—or again, as it is impossible for a chemical reaction to take place unless a certain quantity of matter is present—the phylum cannot establish itself biologically unless, from the start, it has gathered up in itself a sufficient number and variety of potentialities. The lack of a certain initial modicum of consistency and richness (or the failure to break away at a sufficient angle) is enough to prevent a new branch from attaining individuality. The rule is strict. But how, in concrete terms, are we to express the rule and visualise its operation ?—in terms of a diffuse segregation of a mass within a mass, or as an effect of contagion propagating around a narrowly limited area of mutation ? What *surface* representation can we give to the birth of a species ? We are still hesitant and the question may perhaps involve a variety of answers. But we have gone a long way towards solving a problem once we are able to formulate it.

Lastly what serves not only to define the phylum, but also to classify it without ambiguity as one of the *natural units* of the world, is 'its power and singular law of autonomous development'. If we say that it behaves 'like a living thing' this is no mere figure of speech ; in its own way it grows and flourishes.

### B. *The Flourishing of Maturity*

In virtue of analogies which correspond, as we shall discover later, to a deep bond of nature, the development of a phylum is strangely parallel to the successive stages undergone by an invention made by men. We know those stages well from having seen them for about a century constantly around us. Roughly the idea first takes the shape of a theory or a provisional mechanism. Then follows a period of rapid modifications. The rough model is continually touched up and adjusted until it is practically completed. On the attainment of this stage, the new creation enters its phase of expansion and equilibrium. As regards quality it now only undergoes minor changes ; it has reached its ceiling. But quantitatively it spreads out and reaches full consistence. It is the same story with all modern inventions, from the bicycle to thc aeroplane, from photography to the cinema and radio.

In just this way the naturalist sees the curve of growth followed by the branches of life. At the outset the phylum corresponds to the ' discovery ', by groping, of a new type of organism that is both viable and advantageous. But this new type will not attain its most economical or efficient form all at once. For a certain period of time it devotes all its strength, so to speak, to groping about within itself. Try-out follows try-out, without being finally adopted. Then at last perfection comes within sight, and from that moment the rhythm of change slows down. The new invention, having reached the limit of its potentialities, enters its phase of conquest. Stronger now than its less perfected neighbours, the newly born group spreads and at the same time consolidates. It multiplies, but without further diversification. It has now entered its fully grown period and at the same time its period of stability.

The flourishing of the phylum by *simple dilatation* or by the thickening of the initial stalk—except in the case of a branch that has reached the limits of its evolutionary power—this elementary procedure is never completely realised. However decisive and triumphant the solution brought by the new form to the problems raised by existence, it still admits of a certain number of variants. And because each of these variants brings its own particular advantage, they have no power or reason to eliminate each other. That explains why, as it grows, the phylum tends to split up into secondary phyla, each being a variant or 'harmonic' of the fundamental type. It splits up as it were along the whole front of its expansion. It subdivides qualitatively at the same time as it spreads quantitatively. Disjunction starts again. Sometimes the new subdivisions seem merely to correspond to superficial diversifications—they are effects of chance or of a playful inventive exuberance. But at other times they are precise adaptations of the general type to particular needs or habitats. Hence the rays ('radiations') that are clearly marked, as we shall see, in the case of the vertebrates. As is to be expected, the mechanism tends to come into action again, in a more attenuated form, inside each ray. The rays, in their turn, show immediate signs of fanning out in fresh lines of segmentation. Theoretically there is no end to this process. But in fact, as we know by experience, the phenomenon quickly begins to peter out. The process of fanning out soon stops; and the terminal dilatation of the branches goes on without any further appreciable splitting up.

The final picture generally presented by a phylum in full bloom is that of a *verticil of consolidated forms*.

And now—last touch to the whole phenomenon—we find at the heart of each element of the verticil a profound inclination towards socialisation. On the subject of socialisation I must repeat my general observations made above on the vital power of association. Since definite groupings of organised and differentiated individuals or aggregates (ants, bees, mankind) are relatively rare in nature, we might be tempted to think of them as freaks of evolution. But this early impression soon gives way to the opposite conviction—that they exemplify one of the most essential laws of organised matter. Is it the last resort employed by the living group to augment by mutual adherence its resistance to destruction and its capacity for conquest? Is it a useful means for increasing inner wealth by pooling resources ? Whatever the fundamental reason may be, the fact is there : once they have attained their definitive form at the end of each verticillate ray, the elements of a phylum tend to come together and form societies just as surely as the atoms of a solid body tend to crystallise.

Once it has achieved this last progress in consolidating and individualising the extremities of its ramification, the phylum can be said to have attained its full maturity. It will persist, from now on, until it is thinned out and then eliminated either by internal weakening or external competition. Then, except for the accidental survival of a few permanently fixed lines, its story has come to an end—unless by a process of self-fertilisation it starts somewhere or other shooting out a new bud.

To understand the mechanism of this revivification, we must return once again to the idea or symbol of *groping*. As we have already seen, the formation of a verticil is explained in the first place by the phylum's need to pluralise itself in order to cope with a variety of different needs or possibilities. But since the number of stems is always on the increase, and since, moreover, each stem that splits up increases the number of individuals, ' trials ' and ' experiments ' increase in number too. The fanning out of the phylum involves a forest of exploring antennae. And when one of these chances upon the fissure, the formula, giving access to a new compartment of life, then instead of becoming fixed or merely spreading out in monotonous variations, the branch finds all its mobility once more. *It enters on mutation.* Through the new opening, another pulsation of life surges, soon to divide in its turn into verticils under the influence of the combined forces of aggregation and disjunction. A new phylum appears, grows, and spreads out above the branch on which it was born though without necessarily stifling or exhausting it. And so the process continues. Perhaps a third branch germinates on the second, and yet a fourth on the third—always provided the branches are on the right path and the general equilibrium of the biosphere is favourable.

## C. *Effects of Distance*

Thus, by the very rhythm of its development, each line of life follows a process of alternate contraction and expansion. It takes on the appearance of a series of knots and bulges, strung like beads, a sequence of narrow peduncles and spreading leaves.

But this gives only a theoretical representation of what happens. For the process to be *seen* as it really is, we should require a terrestrial witness simultaneously present through the whole of duration, and the very idea is monstrous. In reality, the ascent of life can only be apprehended by us from the standpoint of a short instant, that is through an immense layer of *lapsed* time. What is granted to our experience and which subsequently constitutes the 'phenomenon' is thus not the evolutionary movement in itself; it is this movement corrected according to its alteration by the *effects of distance*. How does this alteration show itself? Quite simply through the accentuation (rapidly increasing with the distance) of the fan-structure deriving from the phyletic radiations of life. This happens, moreover, in two different ways; first by exaggeration of the apparent dispersion of the phyla and subsequently by the apparent suppression of the peduncles.

*Exaggeration of the apparent dispersion of the phyla.* This first optical illusion, affecting all observation, is due to the ageing and to the 'decimation' of the living branches as a result of age. Only an infinitesimal number of the organisms that have grown successively on the tree of life exist for us to inspect today. And, despite all the efforts of palaeontology, many extinct forms will remain unknown to us for ever. As a result of this destruction, many gaps are continually forming in the ramifications of the animal and vegetable kingdom, and the farther back we go, the larger the gaps are. Dried up branches have broken off. Leaves have fallen. Many transitional forms have disappeared and their absence often makes the surviving lines of generation look gaunt and solitary. Duration, which with one hand multiplies its creations ahead, works no less diligently with the other hand thinning out the ranks in the rear. By so doing, it separates them off and isolates them more and more in our vision, while at the same time, by another and more subtle process, it gives us the illusion of seeing them floating like clouds, rootless, over the abyss of past ages.

*Suppression of the peduncles.* Since the heroic times of Lamarck and Darwin, the favourite argument employed against the transformists has always lain in pointing out their incapacity to prove the *birth* of a species in terms of *material traces*. 'Admittedly you show us,' say these objectors, 'a succession of varying forms in past ages, and we will even concede that you are able to demonstrate the transformation of those forms within certain limits. But however primitive it is, your first mammalian is already a mammal, your first equine already a horse, and so on all along the line. Accordingly, though there may well be evolution within a given type, we see no new type produced by evolution.' So the increasingly rare survivors of the 'fixed-type' school still contend.

Quite apart from all the arguments that can be based, as we shall see, on the continual accumulation of palaeontological evidence, there is a more weighty answer (a conclusive proof in fact) with which the 'fixed-type' school's case can be rebutted. It consists in denying the initial assumption. What the anti-transformists are demanding is nothing less than that we should show them the 'peduncle' of a phylum. But this demand is both pointless and unreasonable. To satisfy it we should have to change the very nature of the world and the conditions under which we perceive it.

Nothing is so delicate and fugitive by its very nature as a beginning. As long as a zoological group is young, its characters remain indeterminate, its structure precarious and its dimensions scant. It is composed of relatively few individual units, and these change rapidly. In space as in duration, the peduncle (or, which comes to the same thing, the bud) of a living branch corresponds to a minimum of differentiation, expansion and resistance. What, then, will be the effect of time on this area of weakness ?

Inevitably to destroy all vestiges of it.

Beginnings have an irritating but essential fragility, and one that should be taken to heart by all who occupy themselves with history.

It is the same *in every domain* : when anything really new begins to germinate around us, we cannot distinguish it—for the very good reason that it could only be recognised in the light of what it is going to be. Yet, if, when it has reached full growth, we look back to find its starting point, we only find that the starting point itself is now hidden from our view, destroyed or forgotten. Close as they are to us, where are the first Greeks and Romans ? Where are the first shuttles, chariots or hearth-stones ? And where, even after the shortest lapse of time, are the first motor-cars, aeroplanes or cinemas ? In biology, in civilisation, in linguistics, as in all things, time, like a draughtsman with an eraser, rubs out every weak line in the drawing of life. By a mechanism whose detail in each individual case seems avoidable and accidental, but which, taken over a wide range, expresses a fundamental condition of our knowledge, embryos, peduncles and all early stages of growth fade and vanish as they recede into the past. Except for the fixed maxima, the consolidated achievements, nothing, neither trace nor testimony, subsists of what has gone before. In other words, the terminal enlargements of the fans are only prolonged into the present by their survivors or their fossils.

With that understood, there is nothing surprising in our finding, when we look back, that everything seems to have burst into the world *ready made*.[1] That which moves automatically tends to disappear from our view (by the selective absorption of

[1] If our machines (cars, planes, etc.) were swallowed up in some cataclysm and 'fossilised', future geologists, finding them, would get the same impression as we get from the pterodactyl. Represented only by the latest makes, these products of our invention would seem to them to have been created without any previous evolutionary groping—completed and 'fixed' at the first attempt.

the ages) to become resolved into a discontinuous succession of levels and stabilities throughout the whole domain of what appears to us.[1]

The destructiveness of the past, superimposed on the constructiveness of growth, enables us in the light of science to distinguish and make a diagram of the ramifications of the tree of life.

Let us try to see it in its concrete reality, and to measure it.

## 3. THE TREE OF LIFE

### A. *The Main Lines*

*a. A Quantitative Unit of Evolution : the Layer of the Mammals.* It follows directly from what has gone before that, to get a clear view of the tree of life, we must 'make our eyes see' that part of it only moderately affected by the corrosive action of time. Not too close, or the leaves will get in the way ; not too far, or the branches will lack detail.

Where in nature today can we find such a privileged region ? Undoubtedly in that great family, the mammals.

If mankind constitutes a group which is still 'immature', the mammals form a group which is both adult and *fresh*. Geology provides us with positive evidence of this, and a simple inspection of the internal structure of the group is enough to prove it. Not reaching full florescence until the Tertiary era, their grouping still leaves visible an appreciable number of their most delicate appendices. That is why the kingdom of the mammals has long been and still is the happy hunting-ground for transformist ideas.

[1] I remark later (footnote p. 186) on the subject of 'monogenism' on the non-fortuitous impossibility we find ourselves in (for fortuitous reasons in every case—cf. Cournot) to get beyond a certain limit of precision (of 'separation') in our perception of the very distant past. In all directions (towards the very old and very small, but also towards the very big and very slow) our view is eventually blurred, and outside a certain radius we distinguish nothing at all.

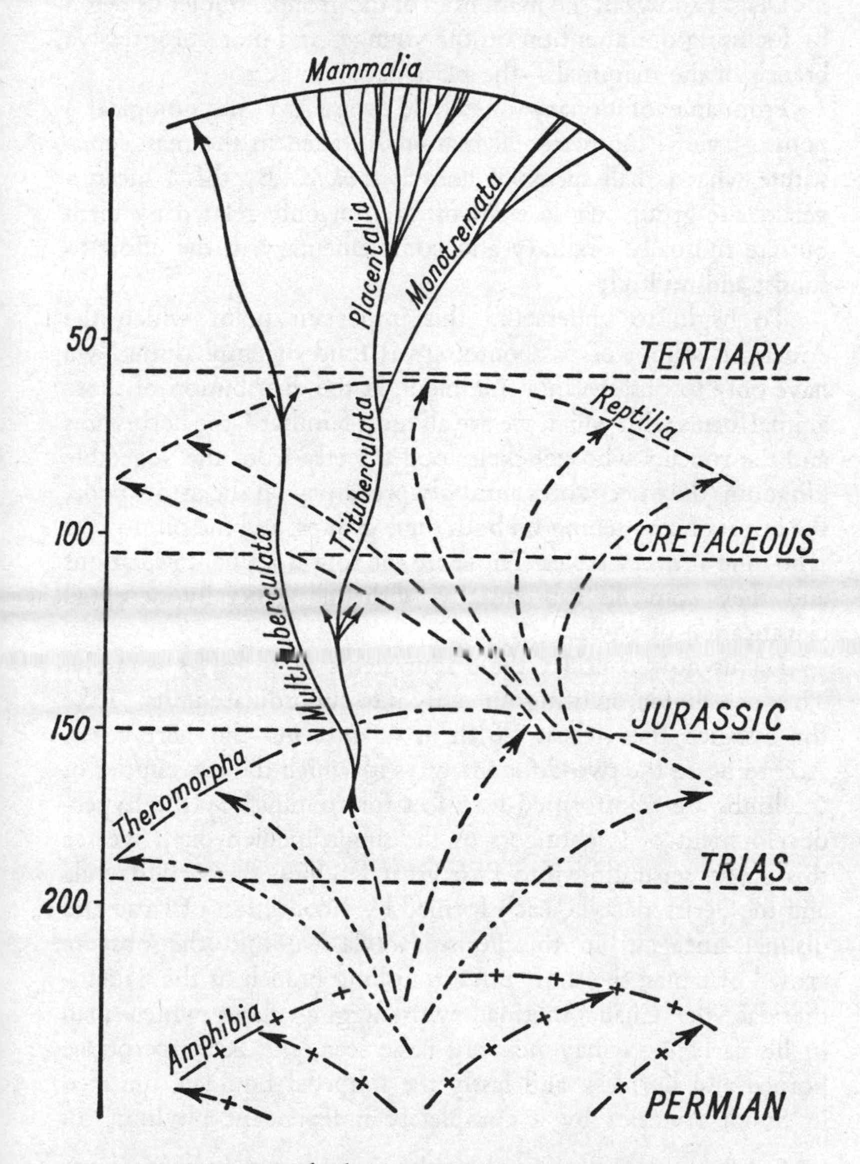


DIAGRAM I. *The development of the Tetrapods in Layers (Birds omitted). The figures on the left indicate millions of years.*

Diag. 1 shows us the main lines of the group. But let us begin by focussing our attention on the younger and more progressive branch of the mammals—the placentals.[1]

From an evolutionary (one could even say a 'physiological') point of view, the placental mammals, taken in the mass, constitute what I shall speak of here as a *biota*. By this I mean a verticillate group whose elements are not only related by birth but are mutually auxiliary and complementary in the effort to subsist and multiply.

To begin to understand this important point which the American school of palaeontology is fond of emphasising, we have only to observe in a suitable light the distribution of those animal forms with which we are all most familiar—the herbivores and the rodents who get their food directly from the vegetable kingdom, the insectivores similarly predatory on the arthropoda, the carnivores battening on both these groups, and the omnivores who dine at every table. Those are the four dominant radiations and they coincide substantially with the generally accepted classification of phyla.

Let us now consider these four stems or sectors separately. They sub-divide, splitting up easily into subordinate units. Take for instance the richest of them at present—the herbivores. According to the two different ways in which the extremities of the limbs are transformed into feet for running (by the hyper-development of two fingers or the single median one), we see this group separating into two great families, the Artiodactyla and the Perissodactyla, each formed by a collection of large and distinct lineages. In the Perissodactyla we find the obscure crowd of tapirs, the short but astonishing branch of the Titanotheridae, the Chalicotheridae with digging claws which man in his early days may possibly have seen, the Rhinocerotidae horned and hornless, and lastly the solipedal Equidae, imitated in South America by a completely independent phylum. In

[1] So called in contrast to the a-placentals (marsupials, etc.) the embryo being nourished by a special organ, the *placenta*, which enables it to develop to maturity in the uterus.

the Artiodactyla we find the Suidae, the Camelidae, the Cervidae and the Antilopidae—to say nothing of other less vigorous stems which are nevertheless as differentiated and as interesting to the palaeontologist. And we have not mentioned that abundant and robust group, the Proboscidia. Conforming to the rule of the 'suppression of the peduncles', the early history of each of these groups is lost in the mists of the past. But once they have appeared we can follow each one of them through the principal phases of their geographical expansion; also through their successive sub-divisions into sub-verticils which proceed almost indefinitely; and lastly by the exaggeration due to orthogenesis of certain skeletal characteristics, dental or cranial, which generally end up by making them monstrous or delicate.

Nor is this all. For we can distinguish, superimposed on this florescence of genera and species issued from the four fundamental radiations, another network corresponding to attempts made here and there to abandon life on the ground and take to the air, the water, or even to an underground existence. Besides forms specialised for running there are arboreal and even flying forms, swimming forms, and burrowing forms. The Cetacea and Sirenia seem to have developed surprisingly quickly from the carnivores and the herbivores. Others (such as the chiroptera, moles and mole-rats) are derived from the oldest elements of the placental group, the insectivores and the rodents both dating from the end of the Secondary era.

One has only to consider this elegantly balanced functional whole to be convinced that it represents an organic and natural grouping which is *sui generis*. This conviction gathers strength when we realise that it does not correspond to an isolated excepttional case, but that similar units have periodically appeared in the course of the history of life. We only need mention two examples within the confines of the mammals.

Geology teaches us that during the Tertiary era a fragment of the placental biota, then in full process of evolution, was cut off by the sea and imprisoned in the southern half of the American continent. Now how did this off-shoot react to its isolation?

Exactly like a plant—that is to say, it reproduced on a smaller scale the same design as the trunk from which it had been separated. It set to work to grow its pseudo-elephants, its pseudo-rodents, its pseudo-horses and its pseudo-monkeys (Platyrrhini). A complete biota in miniature, a sub-biota within the original one.

And now for our second example, furnished by the marsupials.

To judge by their relatively primitive method of reproduction and also by their present geographical distribution (in surviving pockets), the marsupials or a-placentals represent a peculiar stage at the base of the mammalian stem. They must have flourished before the placentals, forming a separate earlier biota of their own. On the whole, except for a few strange types (like that fossil pseudo-Machaerodus recently found in Patagonia),[1] this marsupial biota has disappeared without leaving a trace. On the other hand, one of its sub-biota accidentally developed and conserved in Australia before the Tertiary era and again through isolation, shows such sharpness of contour and perfection as still to make the naturalists marvel. At the time of its discovery by Europeans, Australia, as is well known, was inhabited only by marsupials.[2] They were of great variety, however, being of all shapes, sizes and habitats—herbivorous and cursorial marsupials, carnivorous marsupials, insectivorous marsupials, marsupial rats, marsupial moles, etc. It would be impossible to imagine a more striking example of the power inherent in every phylum to differentiate itself into a sort of closed and physiologically complete organism.

This grasped, let us now lift our eyes to the vast system enclodse by the two biota, the placentals and a-placentals, considered together. Zoologists noticed at an early date that in all the forms composing these two groups, the molar teeth were

[1] *Machaerodus* or sabre-toothed tiger. This big feline, common at the end of the Tertiary era and at the beginning of the Quaternary era, is strangely mimicked by the Pliocene carnivorous marsupial of South America.

[2] Except for a group of rodents and (the latest arrivals) man and his dog.

essentially tritubercular, the projections of upper and lower teeth neatly fitting beside each other ; an insignificant trait in itself, but intriguing because of its constancy. How explain the universality of such an accidental characteristic ? The key to the enigma has been provided by a discovery made in certain Jurassic beds in England. In the Middle Jurassic period, in a flash, we get a glimpse of a first pulsation of mammals—a world of small animals no bigger than rats or shrews. And in these tiny creatures, already extraordinarily varied, the dental type is not yet fixed, as it is in nature at the present day. Among them we can already find the tritubercular type ; but alongside it all sorts of other combinations may be observed in the development and opposition of molars and cusps. These other combinations have been eliminated long since. From this only one conclusion can be drawn. With the possible exception of the Ornithorhynchus and the Echidna (paradoxical oviparous forms sometimes supposed to be a prolongation of the 'multitubercular' type), existing mammals all derive from one narrow unique group. Taken all together they represent (in a state of florescence) *but a single one of the many stems* into which the Jurassic verticil of the mammals was divided—namely the tritubercular.[1]

At this point we have almost reached the limit of what the opacity of the past will allow us to see. Beneath this level, except for the probable existence right at the end of the Triassic period of yet another verticil to which the multitubercular type would seem to belong, the story of the mammals is lost to us.

But at least we can say that towards the top and all round it, their group, naturally isolated by the rupture of its peduncle, stands out with sufficient sharpness and individuality for us to accept it as a *practical unit* of 'evolutionary mass'.

Let us call this unit a *layer*.

We shall be needing that unit at once.

*b. A Layer of Layers : the Tetrapods.* When they measure the

[1] Which might alternatively be called the 'septem-vertebrates' since, by another coincidence which is equally unexpected and significant, all have *seven* cervical vertebrae whatever the length of the neck.

distance of the nebulae, astronomers calculate in light years. If we, working back from the mammals, want to enlarge and prolong our vision of the tree of life downwards, we must calculate in layers.

Let us begin with the layer of the reptiles of the Secondary era.

When we lose sight of it below the Jurassic period, the mammalian branch does not disappear into a sort of vacuum. Instead we find it enveloped and covered over by a thick living growth of an entirely different appearance : of Dinosaurians, Pterosaurians, Ichthyosaurians, Crocodilia and many other monsters less familiar to the layman in palaeontology. Amongst these, the zoological distances between the various forms are considerably greater than between the various orders of mammals. Yet three characteristics strike us at once. Firstly, we are dealing with a ramifying system. Secondly, the ramifications are already far advanced or even nearing the end of their florescence. Thirdly, by and large, the whole group represents nothing else than an immense and perhaps complex biota. The herbivorous forms are often gigantic. Their satellites and enemies, the carnivores, are heavy or leaping types. Besides there are the flying types, with their bat-like membranes or their birds' feathers. Lastly, swimming types, as streamlined as dolphins.

In the distance this reptilian world seems to us more compressed than the mammalian, yet, judged by its expansion and its final complexity, it must be assumed to have lasted at least as long. Anyhow, it disappears into the depths in the same way. About the middle of the Triassic age, Dinosaurians can still be recognised ; but hardly emerging from another layer which itself is approaching its decline, that of the Permian reptiles, best typified in the Theromorphs.

Clumsy and deformed and rare in our museums, the Theromorphs are much less popular than the Diplodocus or the Iguanodons. This does not prevent their taking a position of growing importance on the zoological horizon. At first regarded merely as freaks belonging exclusively to South Africa, they have now been definitively identified as the sole representatives of a complete and special stage in continental vertebrate life. At one moment, before the Dinosaurs, before the mammals, they were the creatures that occupied and possessed all land that was not covered by sea. Standing squarely on their strongly articulated limbs, and often provided with teeth of molar form, they might well be called the first quadrupeds to be firmly established on *terra firma*. In the age in which we become aware of their presence, we find them abounding in a strange variety of forms—horned, crested, armoured—indicating (as always) a group at the end of its evolutionary career. A rather monotonous group, as a matter of fact, under its superficial extravagances. One, moreover, which does not yet exhibit clearly the nervures of a true biota. It is nevertheless a fascinating group by virtue of the spread and the potentialities of its verticil. On the one hand there are the unchangeable tortoises, and at the other extreme, types which in their agility and cranial construction are very progressive. We have every reason to believe that it was among the latter that the long dormant shoot finally appeared which was to become the mammalian branch.

Then another 'tunnel'. At these distances, the slices of duration are increasingly compressed under the weight of the past. When, at the lowest level of the Permian era and below it, we discern another surface of the inhabited earth, we find it now occupied only by amphibians crawling over the slime. The amphibians—a throng of squat or serpentine creatures among which it is often difficult to distinguish adult from larval forms ; skin glabrous or armoured ; vertebrae tubular or in a mosaic of tiny bones. Here again, following the general rule, we can only find an already highly differentiated world, almost coming to an end ; and there may well be many other layers that we confuse in this writhing mass, through sediments about whose thickness and immense duration we are still unclear. But one thing is sure. At this level we are witnessing the emergence of an animal group from the waters in which it was nourished and formed.

And at this extreme beginning of their sub-aerial life, the vertebrates display a surprising characteristic which we must pause to consider. In every variety the skeletal formula is the same, particularly in the number and composition of the locomotory limbs, to say nothing of the marvellous similarities of the cranial bones. What is the reason for this ?

The fact that all amphibians, reptiles and mammals have four legs, and only four, might be explained in terms of mere convergence towards a particularly simple mode of locomotion (though the insects never have less than six legs). But how are we to justify in purely mechanical terms the complete similarity of structure in these four appendices ? In the anterior pair, the single humerus, then the two bones of the forearm and the five digits of the hand. Is not this, yet again, one of those accidental combinations which could *only once* have been discovered and accomplished ? If so, the conclusion already forced upon us in the case of the trituberculariry of mammals looms up again. Despite their extraordinary variety, terrestrial air-breathing animals can only represent variation superposed on a very special solution of life.

Thus when we go back towards its origins, the immense and complicated ramification of the walking vertebrates folds back and closes in upon itself in a single stem.

A single peduncle closes and defines at its base a layer of layers—the world of four-footedness.

*c. The Branch of the Vertebrates.* In the case of mammals, we have been able to pick out the verticil from which the 'tritubercular' stem shot off and isolated itself. Science has made less progress about the origin of the amphibians. We have no hesitation, however, in pointing to the only region of life in which four-footedness could have germinated amongst other tentative combinations. It must have done so somewhere among fish with lobed and 'limb-like' fins whose layer, once widespread, is now represented only by a few 'living fossils'—the Dipnoi (or lung-fishes) and, a very recent surprise, the 'Crossopterygian', recently fished up in the southern seas.

Made superficially 'homogenic' by mechanical adaptation to swimming, the fish (it would be better to call them the Pisciformes) are an assembly of monstrous complexity. We seem to find here more than anywhere numbers of layers accumulated and confused under the same heading. There are relatively young layers developing in the oceans at the very time when those of the four-footed were spreading over the continents. There are also ancient layers, still more numerous, ending up at a very low stage near the Silurian, at a fundamental verticil from which we see two principal stems diverging : Pisciformes with one nostril and no jaws, represented in nature today by the lamprey alone, and Pisciformes with jaws and two nostrils, *from which all the rest have been derived.*

After what I have said above about the concatenation of terrestrial forms, I will no longer attempt to unearth and analyse this other world. I prefer to draw attention to a fact of a different order which we meet here for the first time. The oldest fishes we know are for the most part strongly, even abnormally, scaly.[1] Under this first and apparently rather fruitless attempt at external consolidation was an internal skeleton still entirely cartilaginous. As we go back, the vertebrates appear less and less ossified internally. That is why we lose trace of them, no vestige remaining even in sediments that have come down to us intact. Now this is only a particular example of a general phenomenon of immense importance—whatever living group we take, it always ends by drowning itself in the depths of *mollification.* This is an infallible way of causing peduncles to vanish.

Thus below the Devonian level the Pisciformes disappear into a sort of foetal or larval phase, incapable of fossilisation. Were it not for the accidental survival of the strange Amphioxus, we should have no idea of the multiple stages that the Chordate type had to go through before being ready to fill the waters, pending its invasion of the land.

So at the base a vast vacuum ends the story of that enormous edifice which includes all the quadrupeds and all the fishes, *the branch of the vertebrates.*

[1] Without this ossified integument they would have left nothing behind them and we should never have known of them.

*d. The Remaining Branches of Life.* With the vertebrate branch we have, within the biosphere, the greatest definite group known to systematic biology. Two other branches, and two only, besides the vertebrate, contribute to forming the main ramification of life—one consituted by the worms (Annelida) and arthropods, the other by the vegetable kingdom. The first consolidated by chitin or calcareous matter, the second by cellulose, they, too, succeeded in breaking out of their watery prison, to spread vigorously in the atmosphere. Indeed, in nature today, plants and insects are locked in a struggle with boned animals for the world's available space.

It would be possible to analyse these two other branches as we have just analysed the vertebrates, but I think we can dispense with that. At the top we find the newer groups, rich in delicate verticils ; deeper, the layers with stems more firmly drawn but less well equipped ; and right at the base the fading-away into a world of unstable chemical forms. Thus we see the same general pattern of development ; but because in the latter case the branches are obviously older, there is greater complication and, in the instance of the insects, we observe extreme forms of socialisation.

There seems no reason to doubt that in the abysses of time these various lines converge towards some common pole of dispersion. But long before we reach the junction of the Chordates, the Annelids and the plants (the junction of the first two being among the metazoa, while their junction with the plants is much lower still and among the protozoa), their respective trunks vanish into a complex of extremely strange forms : Porifera, Echinodermata, Coelenterata. All tentative answers to the problem of life, a thicket of abortive branches.

All this emerges beyond question (though we are unable to say how, so wide is the breach of continuity caused by duration) from another world quite unbelievably old and multiform : infusoria, various protozoa and bacteria—free cells, naked or shelled in which the kingdoms of life are confused and which science is unable to classify. Applied to them, the words animal or vegetable lose all meaning. We are no longer able to determine whether we are dealing with layers piled on layers and branches on branches, or a 'mycelium' of confused fibres such as we find in a mushroom. Nor can we say from what all this germinated. Below the Precambrian stage, the unicellular creatures too lose every kind of calcareous or siliceous skeletal form. And so the roots of the tree of life are lost to view in the unknowable world of soft tissue and the metamorphosis of primaeval slime.

### B. *The Dimensions*

So we bring to a close our very sketchy diagram of the forms that have been observed and labelled by the patient labour of naturalists from Aristotle to Linnaeus and onwards. In the course of describing it, we have already tried to communicate the enormous complexity of the world we were attempting to resuscitate. It remains for us, by a final effort of vision and facing it as a whole, to realise more explicitly its prodigious dimensions. Of their own accord our minds always tend, not only to clarify (which is their function) but also to condense and abbreviate the realities they touch. They falter, overburdened by the weight of distances and multitudes. So having sketched, for what it is worth, the expansion of life, it is incumbent on us to restore to the elements of our diagram their true dimensions, in number, in volume and in duration.

Let us now attempt this.

First of all, *in number*, for the sake of simplicity our sketch of the animate world had to be made in bold strokes—families, orders, biota, layers, branches. But in dealing with these collective units, have we really even begun to imagine the multitudes that in fact we were dealing with ? Anyone who wishes to think in terms of evolution, or write about it, should start off by wandering through one of those great museums—there are four or five in the world—in which (at the cost of efforts whose heroism and spiritual value will one day be understood) a host of travellers has succeeded in concentrating in a handful of rooms the entire spectrum of life. There, without bothering about names, let him surrender himself to what he sees around him, and become impregnated by it : by the universe of the insects whose 'reliable' species are counted in tens of thousands ; by the molluscs, thousands more, inexhaustibly variegated in their marblings and their convolutions ; by the fishes, unexpected, capricious, and as prettily marked as butterflies ; by the birds, hardly less extravagant, of every form, feather, and beak ; by the antelopes of every coat, carriage, and diadem. And so on, and so on. And for each word, which brings to our minds a dozen manageable forms, what multiplicity, what impetus, what effervescence ! And to think that all we see are merely the survivors ! What would it be like if all the others were there too ? In every epoch of the earth, on every level of evolution, other museums would have displayed the same teeming luxuriance. Added together, the hundreds of thousands of names in our catalogues do not amount to one millionth of the leaves that have sprouted so far on the tree of life.

Next, *in volume*. By this I mean: what is the relative importance, quantitatively, of the various zoological and botanical groups in nature ? What share belongs to each, materially, in the general assemblage of organised beings ?

To give a rough idea of their proportion, I am reproducing here the very illuminating diagram in which a master in this field, M. Cuénot, has shown the principal departments of the animal kingdom, in the light of the most recent advances in science. This is a diagram of position rather than of structure, but it answers precisely the question I am asking.

Looking at it, we may well receive an initial shock—the sort of shock we get when an astronomer speaks of our solar system as a simple star, of all our stars as a single Milky Way, and of our Milky Way as a mere atom among other galaxies. Mammals—does not that word normally sum up our idea of 'animal' ? Here it is, a poor little lobe, a belated offshoot on the tree of life. Around it, on the other hand, and beneath—

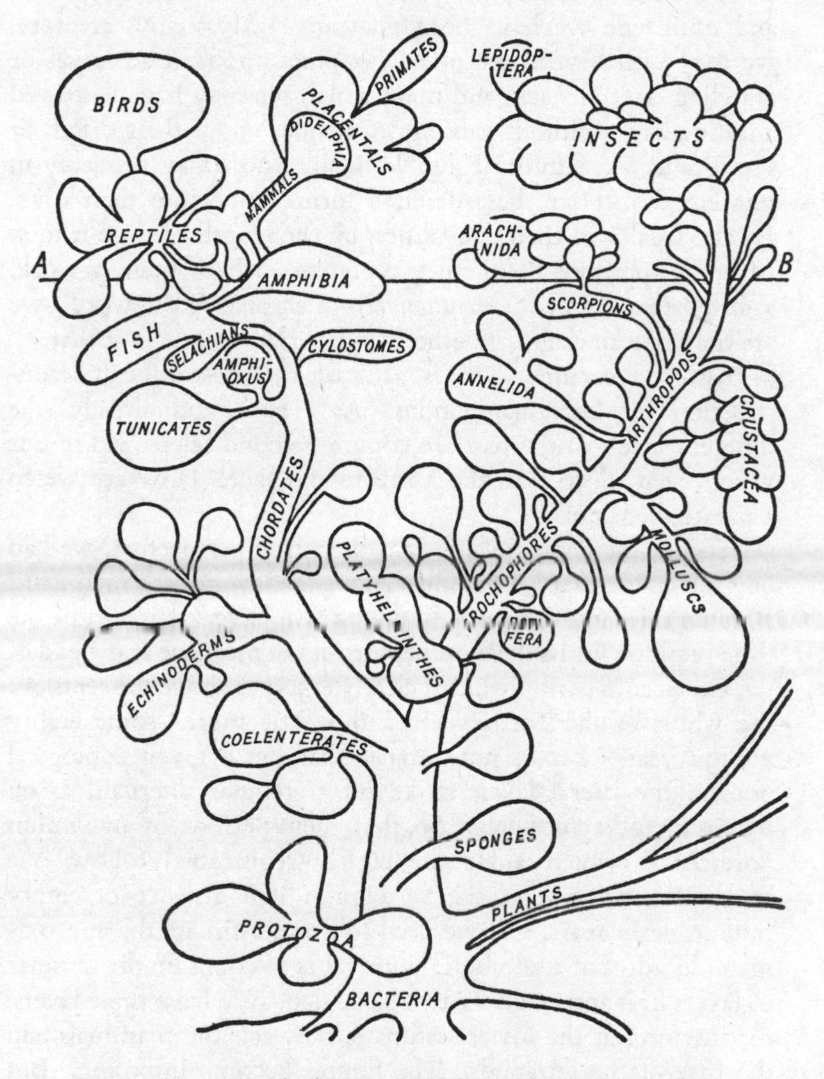


DIAGRAM 2. *The 'Tree of Life' after Cuénot. On this diagram each principal lobe (or bunch) represents a grade at least as important (morphologically and quantitatively) as that of the whole of the Mammalia taken together. Below the line AB, the forms are aquatic; above it they live on land.*

what a teeming rivalry of types, of whose existence, magnitude and multitude we have been unaware ! Mysterious creatures we may well have come upon, hopping among dead leaves or crawling over a beach, and upon which we may have bestowed an idle glance without pausing to wonder about their origin or significance—creatures negligible in size and today probably in number too. Here these despised forms come into their own. By the wealth of their modalities, by the length of time it took nature to produce them, they represent each of them a world as important as ours. *Quantitatively*—I emphasise the word—we are only one among these others, and the latest comers at that.

Lastly, *in duration*. This is, as usual, the most difficult reconstruction for our imaginations. As I have said already, the different levels of the past are compressed and telescoped in our vision even more than the horizons of space. How are we to separate them out ?

To put the depths of life into their truc perspective, we had best return to what I have called above the layer of the mammals. Because this layer is relatively young, we have some idea of the time required for its development from the moment at the end of the Cretaceous period when it clearly emerges above the reptiles : the whole of the Tertiary era and a little more—some eighty million years. Let us now *assume* that, on a given zoological branch, the lateral layers strike off at regular intervals, as on the trunk of a pine-tree. So that their periods of maximum florescence (which alone are clearly registrable) follow one another in the case of the vertebrates at a distance of eighty million years apart. All we need to do to estimate the approximate duration of a zoological interval is to count up the number of layers in it and multiply by 80,000,000. We have three layers, for instance, at the lowest estimate, between the mammals and the base of the tetrapods. The figures become imposing. But they tally well enough with current geological ideas as to the immensity of the Triassic, Permian and Carboniferous ages.

We can try to follow another method in a more approximate way from branch to branch. Within one and the same layer—

such as the mammals, once again—we can apprehend vaguely the average structural divergence of types, a divergence which, as we have seen, took some eighty million years. Now compare the mammals, the insects and the higher plants ; unless (which is possible) the three branches at whose ends these three groups flourish did not strike off exactly from the same stem but shot up separately from a common 'mycelium'. What length of time was necessary to effect the gigantic divergence we see ? Here the zoologist's figures would seem as if tending to contradict the geologist's. Physicists, having measured the lead-content of a radiferous Precambrian mineral, are prepared to allow only fifteen hundred million years from the earliest sediment of carbon onwards. Must not the first organisms have existed long before these first vestiges ? Besides, if there is disagreement, which of the two time-measurements shall we trust to count the years of the earth ? The slow disintegration of radium or the slow aggregation of living matter ?

If it takes five thousand years for a mere sequoia to reach its full growth (and no one yet has seen one die a natural death) what can be the total age of the tree of life ?

### C. *The Evidence*

Now we can see the tree of life standing before us. A strange tree, no doubt. We could call it the negative of a tree, for contrary to what happens with our great forest trees, its branches and trunk are revealed to our eyes only by ever-widening gaps ; an almost petrified tree, as it appears to us, so long do the buds take to open. Many that are half-opened now we shall never know in any other state. A clearly drawn tree, none the less, with its superimposed foliage of living species. In its main lines and vast dimensions, it stands there before us covering all the earth. Before attempting to probe the secret of its life, let us take a good look at it. For, from a merely external contemplation of it, there is a lesson and a force to be drawn from it : *the sense of its testimony*.

We still find here and there in the world people whose minds are suspicious and sceptical as regards evolution. Having only a book-knowledge of nature and naturalists these people imagine that the transformist battle is still carried on as in the days of Darwin. And because biologists continue to discuss the mechanisms by which species could have been formed, they imagine that biologists hesitate (or that they could hesitate without suicide) about the fact and reality of such a development.

But the real situation is quite otherwise.

In the course of this chapter devoted to the concatenations of the organised world, the reader may have been surprised at my failing so far to mention the still lively quarrels over the distinction between the 'soma' and the 'germplasm', over the existence and function of 'genes', over the transmission or non-transmission of acquired characters. The truth is that at the point I have reached in my inquiry, these questions do not concern me directly. To provide anthropogenesis with a natural framework and man with a cradle—to guarantee, I mean, the substantial objectivity of an evolution—one thing, and one thing only, is necessary. Namely that the general phylogenesis of life (whatever the process and springboard of it may be) should be as clearly recognisable as the individual orthogenesis through which we see without the least astonishment every living creature pass.

Now a quasi-mechanical proof of this global growth of the biosphere imposes itself inescapably on our minds by the material pattern at which we inevitably end up with each new effort to fix, point by point, the contours and nervures of the organised world.

No one would think of doubting the gyratory origin of the spiral nebulae, the progressive accretion of particles at the heart of a crystal or of a stalagmite or the concretion of the woody 'bundles' round the axis of a stalk. Certain geometrical disposition, which seem to us perfectly stable, are the trace and irrefutable sign of kinematics. How could we hesitate even for a moment about the evolutionary origins of the layer of life on the earth ?

Under our efforts at analysis life sheds its husk. It breaks down to an infinite degree into an anatomically and physiologically coherent system of overlapping fans.[1] We find barely appreciable fans of sub-species and races ; larger ones of species and genera ; still larger ones of biota, then of layers, then of branches. And, to end with, the whole assemblage, animal and vegetable, forming by association one single gigantic biota, rooted perhaps, like a simple stem, in some verticil steeped in the depths of the mega-molecular world. Life would thus be a simple branch based on something else.

From top to bottom, from the biggest to the smallest, one same visible structure whose design, reinforced by the very disposition of the shadows and voids, is accentuated and prolonged (no hypothesis this!) by the quasi-spontaneous arrangement of the unforeseen elements brought forth by the day. Each newly-discovered form finds its natural place, though of course nothing within the framework is absolutely 'new'. What more do we need to be convinced that all this was *born*, that all this has *grown* ?

Thenceforward we can go on for years arguing about the way in which the enormous organism could have come into being. As we look closer at the bewildering complexity of the mechanism, our brains begin to reel. How are we to reconcile this persistent growth with the determinism of the molecules, the blind play of the chromosomes, the apparent incapacity to transmit individual acquisitions by generation ? How, in other words, are we to reconcile the external, 'finalist' evolution of

[1] As regards these fans, it would of course be possible to trace the connections in another way, especially in giving more importance to the parallelisms and convergence. The tetrapods, for example, could be regarded as a bundle composed of several stems derived from different verticils, each one having arrived similarly at the quadruped formula. This polyphyletic scheme fits the facts less well, in my opinion. In any case its truth would not in the least affect my fundamental thesis, viz. that life displays an organically articulated unity which manifestly indicates the phenomenon of growth.

*phenotypes* with the internal, mechanistic evolution of *genotypes* ? Though we take it apart, we still cannot understand how the machine works. This may well be, but the machine is meanwhile standing in front of us ; and it works all the same. Because chemistry is still floundering over the formation of granites, should we dispute the fact that the continents become more granitic year by year ?

*Like all things* in a universe in which time is definitely established as a *fourth dimension*, life is, and only can be, a reality of evolutionary nature and dimension. Physically and historically it corresponds with a function X which determines the position of every living thing in space, in duration and in form.

This is the fundamental fact which requires an explanation : but the *evidence* for it is henceforward above all verification, as well as being immune from any subsequent contradiction by experience.

At this degree of generalisation, it may be said that the problem of transformism no longer exists. The question is settled once and for all. To shake our belief now in the reality of biogenesis, it would be necessary to uproot the tree of life and undermine the entire structure of the world.[1]

[1] As a matter of fact, in view of the impossibility of empirically perceiving any entity, animate or inanimate, otherwise than as engaged in the time-space series, evolutionary theory has long since ceased to be a hypothesis, to become a (dimensional) condition which all hypotheses of physics or biology must henceforth satisfy. Biologists and palaeontologists are still arguing today about the way things happen, and above all about the mechanism of life's transformations, and whether there is a preponderance of chance (the Neo-Darwinians) or of invention (the Neo-Lamarckians) in the emergence of new characters. But on the general and fundamental fact that organic evolution exists, applicable equally to life as a whole or to any given living creature in particular, all scientists are today in agreement for the very good reason that they couldn't practise science if they thought otherwise. The one regret we might express here (and not without astonishment) is that despite the clearness of the facts, this unanimity does not go so far as to admit the 'galaxy' of living forms constitutes (as posited in these pages) a vast 'orthogenetic' movement of involution on an ever-greater complexity and consciouness. But we shall return to this at the conclusion of this book.

## CHAPTER THREE

# DEMETER

THROUGHOUT THE foregoing chapter we spoke of growth to express life's way of proceeding. We were even able to go some way towards recognising the principle behind this impetus which seemed to us linked up with the phenomenon of *controlled additivity*. By a continuous accumulation of properties (whatever the exact hereditary mechanism involved) life acts like a snowball. It piles characters upon characters in its protoplasm. It becomes more and more complex. But, taken as a whole, what is the meaning of this movement of expansion ? Is it like the confined and functional explosion of the internal combustion engine ? Or is it a disorderly release of energy in all directions like the blast of a high explosive ?

That there is *an* evolution of one sort or another is now, as I have said, common ground among scientists. Whether or not that evolution is *directed* is another question. Asked whether life is *going anywhere* at the end of its transformations, nine biologists out of ten will today say no, even passionately. They will say : 'It is abundantly clear to every eye that organic matter is in a state of continual metamorphosis, and even that this metamorphosis brings it with time towards more and more improbable forms. But what scale can we find to assess the absolute or even relative value of these fragile constructions ? By what right, for instance, can we say that a mammal, or even man, is more advanced, more perfect, than a bee or a rose ? To some extent we can arrange beings in increasingly wide circles according to the distance in time which separates them from the initial cell. But, once a certain degree of differentiation has been reached, we can no longer find any scientific grounds for preferring one of these laborious products of nature to another. They are different solutions—but each equivalent to the next. One spoke of the wheel is as good as any other ; no one of the lines appears to lead anywhere in particular.'

Science in its development—and even, as I shall show, mankind in its march—is marking time at this moment, because men's minds are reluctant to recognise that evolution has a precise *orientation* and a privileged *axis*. Weakened by this fundamental doubt, the forces of research are scattered, and there is no determination to build the earth.

Leaving aside all anthropocentrism and anthropomorphism, I believe I can see a direction and a line of progress for life, a line and a direction which are in fact so well marked that I am convinced their reality will be universally admitted by the science of tomorrow. And I want here to make the reader understand why.

## 1. ARIADNE'S THREAD

To begin with, as we are dealing here with degrees of organic complication, let us try to find an order in the complexity.

Contemplated without any guiding thread, it must be recognised that the host of living creatures forms qualitatively an inextricable labyrinth. What is happening, where are we going through this monotonous succession of ramifications ? In the course of ages, doubtless, creatures acquire more organs of increased sensibility. But they also reduce them by specialisation. Besides, what is the real meaning of the term 'complication' ? There are so many different ways in which an animal can become less simple—differentiation of limbs, of tissues, of sensory organs, of integument. According to the point of view adopted, all sorts of distributions are possible. In these multiple combinations, is there really one which can be said to be *truer* than the others ? Is there one, that is to say, which gives to the whole of living things a more satisfying coherence, either in relation to itself, or in relation to the world to which life finds itself committed ?

To answer this question, I think we had better go back to what I said above about the mutual relations between the *without* and the *within* of things. The essence of the real, I said, could well be represented by the 'interiority' contained by the universe at a given moment. In that case evolution would fundamentally be nothing else than the continual growth of this 'psychic' or 'radial' energy, in the course of duration, beneath and within the mechanical energy I called 'tangential', which is practically constant on the scale of our observations (Book I, Chapter 2, 3 Spiritual Energy, Section B). And what, I asked, is the particular co-efficient which empirically expresses the relationship between the radial and tangential energies of the world in the course of their respective developments ? Obviously *arrangement*, the arrangement whose successive advances are inwardly reinforced, as we can see, by a continual expansion and deepening of consciousness.

Let us turn this proposition round (not in a vicious circle, but by a simple adjustment of perspective). Among the innumerable complications undergone by organic matter in ebullience, we find it hard to distinguish those which are merely superficial diversifications and those (if any) which would represent a renewal and re-grouping of the stuff of the universe. Well then, let us just try to see whether, amongst all the combinations tried out by life, some are not organically associated with a positive variation in the psychism of those beings which possess it. If so, let us seize on them and follow them ; for, if my hypothesis be correct, they are undoubtedly the ones which, among the equivocal mass of insignificant transformations, represent the very essence of complexity, of essential metamorphosis. There is every chance that they will lead us somewhere.

Framed in these terms, the problem is immediately solved. Of course there exists in living organisms a selective mechanism for the play of consciousness. We have merely to look into ourselves to perceive it—the nervous system. We are in a positive way aware of one single 'interiority' in the world: our own directly, and at the same time that of other men by immediate equivalence, thanks to language. But we have every reason to think that in animals too a certain inwardness exists, approximately proportional to the development of their brains. So let us attempt to classify living beings by their degree of 'cerebralisation'. What happens? An order appears—the very order we wanted—and automatically.

To begin with, let us turn to that part of the tree of life we know best, partly because it is still full of vitality and partly because we belong to it ourselves—the Chordate branch. In this group an outstanding characteristic is apparent, one which has for long been emphasised in palaeontology. It is that we find from layer to layer, by massive leaps, the nervous system continually developing and concentrating. We all know the example of the enormous Dinosaurs whose absurdly small brain was no more than a narrow string of lobes considerably smaller in diameter than the spinal chord in the lumbar region, reminding us of the state of affairs still lower, in the amphibians and the fishes. But when we pass to the stage above—the mammals—we see a remarkable change.

Among the mammals, that is to say, this time, *within a single layer*, the average brain is much more voluminous and convoluted than in any other group of vertebrates. Yet, when we look closer, we see not only many inequalities, but a remarkable order in their distribution. The gradation in the first place follows the position of the biota. In nature at the present day the placentals take precedence in the matter of brain over the marsupials. Next, within the same biotas, we find a gradation according to age. We see placental brains (except for a few primates) always relatively smaller and simpler in the lower Tertiary age than in the Miocene and Pliocene. This is strongly emphasised by extinct phyla such as the Condylarthra or Dinocerata, those horned monsters whose brain-case (in size and the spacing of the lobes) had hardly advanced beyond that of the Secondary reptiles. This can also be observed *within a single line of descent.* In the Eocene carnivores, for instance, the cerebrum, still in the marsupial stage, is smooth and well separated from the cerebellum. It would be easy to add to the list. In general it may be said that, taking any offshoot from any verticil, it is only rarely that we find that (provided it is long enough) it does not lead in time to more and more 'cerebralised' forms.

Taking another branch, the arthropods and the insects, we find the same phenomenon. Dealing as we are now with another sort of consciousness, we are less sure of our values, but the thread which guides still seems to hold. From group to group and age to age, these forms, psychologically so far removed, display, like ourselves, the influence of cerebralisation. The nerve ganglions concentrate ; they become localised and grow forward in the head. At the same time instincts become more complex ; and simultaneously the extraordinary phenomena of socialisation appear, to which we shall have to return.

We could continue this analysis indefinitely. I have said enough, however, to show how easily the skein is disentangled once we have found the end. For obvious reasons of convenience, naturalists setting out to classify organic forms have been led to make use of certain variations of ornament, for instance, or functional modifications of the skeleton. Guided by orthogenesis affecting the coloration and nervation of wings, the disposition of limbs, or the shape of teeth, their classification sorts out the fragments or even the skeleton of a structure in the living world. But because the lines thus traced correspond only to the secondary harmonics of evolution, the system as a whole has neither shape nor movement. On the other hand, from the moment that the measure (or parameter) of the evolving phenomenon is sought in the elaboration of the nervous systems, not only do the countless genera and species fall naturally into place, but the entire network of their verticils, their layers, their branches, rises up like a quivering spray of foliage. Not only does the arrangement of animal forms according to their degree of cerebralisation correspond exactly to the classification of systematic biology, but it also confers on the tree of life a sharpness of feature, an impetus, which is incontestably the hall-mark of truth. Such coherence—and, let me add, such ease, inexhaustible fidelity and evocative power in this coherence—could not be the result of chance.

Among the infinite modalities in which the complication of life is dispersed, the differentiation of nervous tissue stands out, as theory would lead us to expect, as a significant transformation. *It provides a direction ; and therefore it proves that evolution has a direction.*

That is my first conclusion. But it has its corollary. We began by saying that, among living creatures, the brain was the sign and measure of consciousness. We have now added that, among living creatures, the brain is continually perfecting itself with time, so much so that a given quality of brain appears essentially linked with a given phase of duration.

The final conclusion proclaims itself, a conclusion which at one and the same time confirms the bases and controls what follows in our disquisition. Since, in its totality and throughout the length of each stem, the natural history of living creatures amounts on the *exterior* to the gradual establishment of a vast nervous system, it therefore corresponds on the *interior* to the installation of a psychic state coextensive with the earth. On the surface, we find the nerve fibres and ganglions ; deep down, consciousness. We were only looking for a simple rule to sort out the tangle of appearances. And now (entirely in keeping with our initial anticipations on the ultimately psychic nature of evolution) we possess a fundamental variable capable of following in the past, and perhaps defining in the future, the true curve of the phenomenon.

Will that solve the problem ? Yes, almost. But on one condition, obviously ; a condition which will seem irksome to certain scientific prejudice. It is that by a change of front, a reversal of plane, we abandon the *without* to delve into the *within of things.*

## 2. THE RISE OF CONSCIOUSNESS

Let us return to the ' expansionist ' movement of life as it appears in its broad outline. But this time, instead of losing ourselves in the labyrinth of arrangements affecting the ' tangential ' energies of the world, let us try to follow the ' radial ' progress of its internal energies. Now everything becomes definitively clear—in value, in operation and in hope.

*a.* To begin with, what is brought to light by this simple change of variable is *the place occupied by the development of life in the general history of our planet.*

When we discussed the origin of the first cells, we considered that, if their spontaneous generation took place only once in the whole of time, it was apparently because the initial formation of the protoplasm was bound up with a state which the general chemistry of the earth passed through only once. The earth, we said, should be regarded as the seat of a certain global and irreversible evolution, much more important for scientists to consider than any superficial oscillations. We said, moreover, that the primordial emergence of organised matter marked a critical point on the curve of this evolution.

After that the phenomenon seemed to become lost in the multitude of ramifications, to the point that we almost forgot it. But now we see it emerging again, on the tide, with the tide (duly recorded by the nervous systems), whose flood carries the ving mass ever onward towards more consciousness. This is the great primaeval movement reappearing, whose sequel we now grasp.

Like the geologist occupied in recording the movements of the earth, the faultings and foldings, the palaeontologist who fixes the position of the animal forms in time is apt to see in the past nothing but a monotonous series of homogeneous pulsations. In these records, the mammals succeeded the reptiles which succeeded the amphibians, just as the Alps replaced the Cimmerian Mountains which had in their turn replaced the Hercynian range.

Henceforward we can and must break away from this view which lacks depth. We have no longer the crawling 'sine' curve, but the spiral which springs upward as it turns. From one zoological layer to another, *something is carried over: it grows, jerkily, but ceaselessly and in a constant direction.* And this 'something' is what is most physically essential in the planet we live on. The evolution of the simple bodies following the radio-active way, the granitic segregation of continents, the possible isolation of the interior layers of the globe many other transformations besides the vital movement form no doubt a continuous bass underlying the rhythms of the earth ; but since life separated out within the heart of matter, these various processes have no longer the quality of being the supreme event. With the birth of the first albuminoids, the essence of the terrestrial phenomenon shifted in a decisive way to become concentrated in that seemingly negligible thickness, the biosphere. The axis of geogenesis is now extended in biogenesis, which in the end will express itself in psychogenesis.

From an inward point of view, constantly confirmed by ever-increasing harmonies, the different objects of science become visible in proper perspective and in their true proportions. We see life at the head, with all physics subordinate to it. And at the heart of life, explaining its progression, the impetus of a rise of consciousness.

*b. The Impetus of Life.* This is a question hotly debated by naturalists ever since the understanding of nature has been hinged on the understanding of evolution. Faithful to their analytical and determinist methods, biolgists persist in looking for the principle of vital developments in external stimuli or in statistics: the struggle for survival, natural selection and so on. From this point of view, the animate world could never advance—if it advanced at all—otherwise than by the automatically regulated sum of all the efforts it makes to remain itself.

Far be it from me, let me say once again, to deny the important, indeed essential, role, played by this historic working of the material forms. As living beings, we feel it in ourselves. To jolt the individual out of his natural laziness and the rut of habit, and also from time to time to break up the collective frameworks in which he is imprisoned, it is indispensable that he should be shaken and prodded from outside. What would we do without our enemies ? While capable of supply regulating within organic bodies the blind movement of molecules, life seems still to exploit for its creative arrangements the vast reactions which are born fortuitously throughout the world between material currents and animate masses. Life seems to play as cleverly with collectivities and events as with atoms. But what could this ingeniousness and these stimulants do if applied to a fundamental inertia ? And what, moreover, as we have pointed out, would the mechanical energies themselves be without some *within* to feed them ? Beneath the 'tangential' we find the 'radial'. The impetus of the world, glimpsed in the great drive of consciousness, can only have its ultimate source in some *inner* principle, which alone could explain its irreversible advance towards higher psychisms.

How can life respect determinism on the *without* and yet act in freedom *within* ? Perhaps we shall understand that better some day.

Meanwhile the vital phenomenon seems on the whole both natural and possible when once the reality of a fundamental impetus has been accepted. Furthermore, its micro-structure itself becomes clearer. For we now perceive a new way of explaining, over and above the main stream of biological evolution, the progress and particular disposition of its various phyla.[1]

[1] In various quarters I shall be accused of showing too Lamarckian a bent in the explanations which follow, of giving an exaggerated influence to the *Within* in the organic arrangement of bodies. But be pleased to remember that, in the 'morphogenetic' action of instinct as here understood, an essential part is left to the Darwinian play of external forces and to chance. It is only really through strokes of chance that life proceeds, but strokes of chance which are recognised and grasped—that is to say, psychically selected. Properly understood the 'anti-chance' of the Neo-Lamarckian is not the mere negation of Darwinian chance. On the contrary it appears as its utilisation. There is a functional complementariness between the two factors ; we could call it 'symbiosis'.

It is one thing to notice that in a given line in the animal kingdom limbs become solipedal or teeth carnivorous, and quite another to guess how this tendency was produced. It is all very well to say that a mutation occurs at the point where the stem leaves the verticil. But what then ? The later modifications of the phylum are as a rule so gradual, and so stable are sometimes the organs affected, even from the embryo (the teeth, for example), that we are definitely forced to abandon the idea of explaining every case simply as the survival of the fittest, or as a mechanical adaptation to environment and us. So what follows ?

The more often I come across this problem and the longer I pore over it, the more firmly is it impressed upon me that in fact we are confronted with an effect not of external forces but of psychology. According to current thought, an animal develops its carnivorous instincts *because* its molars become cutting and its claws sharp. Should we not turn the proposition around ? In other words if the tiger elongates its fangs and sharpens its claws is it not rather because, following its line of descent, it receives, develops and hands on the 'soul of a carnivore' ? It is the same with the timid cursorial types, the same with those that burrow, swim or fly. There is an evolution of characters certainly ; but on condition that this word is taken in the sense of 'temperament'. At first sight the explanation reminds one of the 'virtues' of the Schoolmen. As we go deeper, it becomes increasingly likely. In the individual, qualities and defects develop with age. Why (or rather, how) should they

---

It may be added that if we give its proper place to the essential distinction (still too often ignored) between a biology of small units and a biology of big complexes—in the same way as there is a physics of the infinitesimal and another of the immense—we appreciate the advisability of distinguishing two major zones of the organic world, and treating them differently. On the one hand is the Lamarckian zone of very big complexes (above all, man) in which anti-chance can be seen to dominate ; on the other hand the Darwinian zone of small complexes, lower forms of life, in which anti-chance is so swamped by chance that it can only be appreciated by reasoning and conjecture, that is to say indirectly. (See p. 302.)

not be accentuated phyletically ? And why, on that scale, should they not react upon the organism to stamp it with their image ? After all the ants and termites succeed in fitting out their warriors and their workers with an exterior suited to their instincts. And we also surely know men of prey ?

*c.* Once we have admitted this, unexpected horizons rise up in front of biology. For obvious practical reasons we are led to make use of the variations in their fossilisable parts to follow the links between living creatures. But this practical necessity must not be allowed to blind us to what is limited and superficial in this arrangement. The number of bones, shape of teeth, ornamentation of the integument—all these 'visible characters' form merely the outward garment round something deeper which supports it. We are dealing with only one event, the grand orthogenesis of everything living towards a higher degree of immanent spontaneity. Secondarily, we find by periodical dispersal of this impetus, the verticil of the little orthogeneses, where the fundamental current splits up to form the true, inner axis of each 'radiation'. Finally, thrown over all that like a simple sheath, we find the veil of tissues and the architecture of the limbs. That is the situation.

To write the true natural history of the world, we should need to be able to follow it from *within*. It would thus appear no longer as an interlocking succession of structural types replacing one another, but as an ascension of inner sap spreading out in a forest of consolidated instincts. Right at its base, the living world is constituted by consciousness clothed in flesh and bone. From the biosphere to the species is nothing but an immense ramification of psychism seeking for itself through different forms. That is where Ariadne's thread leads us if we follow it to the end.

In the present state of our knowledge, of course, we cannot dream of expressing the mechanism of evolution in this 'interiorised', 'radial' form. On the other hand, one thing becomes clear. It is that, if this is the real significance of transformism, life, in so far as it represents a *controlled* process, could only proceed ever farther along its original line on condition that it underwent some profound readjustment at a given moment.

The law is formal. We referred to it before, when we spoke of the birth of life. No reality in the world can go on increasing without sooner or later reaching a critical point involving some change of state. There is a ceiling limit to speeds and temperatures. If we increase the acceleration of a body until we get near the speed of light, it acquires by excess of mass an infinitely inert nature. If we heat it, it would first melt, then vaporise. And the same applies to all known physical properties. So long as we could regard evolution as a simple advance towards complexity, we could imagine it developing indefinitely in its own likeness ; there is no ceiling limit to pure diversification. Now that, beneath the historically increasing intricacy of forms and organs, we have discovered the irreversible increase, not only in quantity but also in *quality*, of brains (and therefore consciousness) we are forced to realise that an event of another order—a *metamorphosis*—was inevitably awaited to wind up this long period of synthesis in the course of geological time.

We must now turn our attention to the first symptoms of this great terrestrial phenomenon which ends up in man.

## 3. THE APPROACH OF TIME

Let us return to the wave of life in movement where we left it, i.e. at the expansion of the mammals or, to situate ourselves concretely in duration, let us go back to the world as we can imagine it towards the end of the Tertiary period.

A great calm seems to be reigning on the surface of the earth at this time. From South Africa to South America, across Europe and Asia, are fertile steppes and dense forests. Then other steppes and other forests. And amongst this endless verdure are myriads of antelopes and zebras, a variety of proboscidians in herds, deer with every kind of antler, tigers, wolves, foxes and badgers, all similar to those we have today. In short, the landscape is not too dissimilar from that which we are today seeking to preserve in National Parks—on the Zambesi, in the Congo, or in Arizona. Except for a few lingering archaic forms, so familiar is this scene that we have to make an effort to realise that *nowhere* is there so much as a wisp of smoke rising from camp or village.

It is a period of calm profusion. The mammalian layer has spread out. Yet evolution cannot be stopped. Something, somewhere, is unquestionably accumulating and ready to rise up for another forward leap. But what ? and where ?

To detect what at this moment is maturing in the womb of the universal mother, let us make use of the index which we have henceforward at our disposal. Life is the rise of consciousness, we have agreed. If it is to progress still further it can only be because, here and there, the internal energy is secretly rising up under the mantle of the flowering earth. Here and there, within nervous systems, psychic tension is doubtless increasing. Physicists and doctors use delicate instruments on bodies : let us do likewise, applying our 'thermometer' of consciousness to this somnolent nature. In what region of the biosphere in the Pliocene period is there a sign of rising temperature ?

Of course we must look at heads.

Outside the vegetable kingdom, which does not count,[1] there are two summits of branches, and only two, which emerge before us in air, light and spontaneity : on the arthropod side, the insects ; on the vertebrate side, the mammals. To which side belongs the future—and truth ?

*a. The Insects.* In the higher insects a cephalic concentration of nerve ganglions goes hand in hand with an extraordinary wealth

[1] In the sense that in the vegetable kingdom we are unable to follow along a nervous system the evolution of a psychism obviously remaining diffuse. That is not to say that the latter does not exist, growing in its own manner. I would not think of denying it. Indeed, to take one example out of a thousand, is it not enough to see how certain plants trap insects to be convinced that the vegetable branch, albeit from afar, is like the other two, subservient to the rise of consciousness.

and precision of behaviour. We cannot but wonder when we see living around us this world so marvellously adjusted and yet so terribly far away. Our rivals ? Our successors, perhaps ? Must we not rather say a multitude pathetically involved and struggling in a blind alley ?

What seems to eliminate the hypothesis that the insects represent the issue—or even that they simply are *an* issue—for evolution is the fact that although very much the elders of the higher vertebrates by the date of their florescence, and now they seem irremediably ' stationary '. Throughout what may well be geological ages, they have become endlessly complicated like Chinese characters, yet give the impression of being unable to change their plan—as if their impetus or fundamental metamorphosis were stopped. And if we reflect a moment, we can see certain reasons for this marking-time.

First of all insects are too small. For quantitative development of the organs, an external, chitinous skeleton is a bad solution. In spite of repeated moultings it imprisons the organs : and it quickly yields under increasing interior volumes. The insect cannot grow beyond an inch or two without becoming dangerously fragile. In spite of the disdain with which we sometimes regard ' a mere question of size ', it is undeniable that certain qualities, *by the very fact that they are linked to a material synthesis*, are only capable of being manifested above certain quantities. The superior psychic levels demand physically big brains.

And then, precisely perhaps for this very reason of size, insects show a strange psychic inferiority in the very domain where we should have been tempted to put their superiority. Our own cleverness is dumbfounded by the precision of their movements and their constructions. But we must be careful. Looked at more closely, this perfection is conditioned by the extreme rapidity with which their psychology becomes mechanised and hardened. It has been amply demonstrated that the insect disposes of an appreciable margin of indetermination and choice for its operations. Only, hardly are these performed, than its acts seem to become charged with habit and soon transformed into organic reflexes. Automatically and continually, one could say, its consciousness is extraverted to become frozen at once : (i) in its behaviour, which successive corrections promptly registered render ever more precise and (ii) in the long run, in a somatic morphology in which individual particularities disappear, absorbed by function. Hence those adjustments of organs and behaviour at which Fabre rightly marvelled, and hence also the simply prodigious arrangements which group together in a single living machine the swarming hive or ant-hill.

This could be called a paroxysm of consciousness, which spreads outwards from within, to become materialised in rigid arrangements. The exact opposite of a concentration.

*b. The Mammals.* Let us therefore leave the insects and return to the mammals.

At once we feel at ease ; so much at ease that our relief could be accounted for by an impression of ' anthropocentrism '. If we breathe more freely now that we have come away from the hive and ant-hill, is it not quite simply because, amongst the higher vertebrates, we feel ' at home ' ? There is always the menace of relativity hanging over our minds.

No, we are not making a mistake. In this case at least we are not misled by an impression—our judgment is really being guided by our intelligence, with the power it has to appreciate certain absolute values. If a furry quadruped seems so ' animated ' compared with an ant, so genuinely alive, it is not only because of a zoological kinship we have with it. In the behaviour of a cat, a dog, a dolphin, there is such suppleness, such unexpectedness, such exuberance of life and curiosity ! Instinct is no longer narrowly canalised, as in the spider or the bee, paralysed in a single function. Individually and socially it remains flexible. It takes interest, it flutters, it plays. We are dealing with an entirely different form of instinct in fact, and one not subject to *the limitations imposed upon the tool by the precision it has attained.* Unlike the insect, the mammal is no longer completely the slave of the phylum it belongs to. Around it an ' aura ' of freedom begins

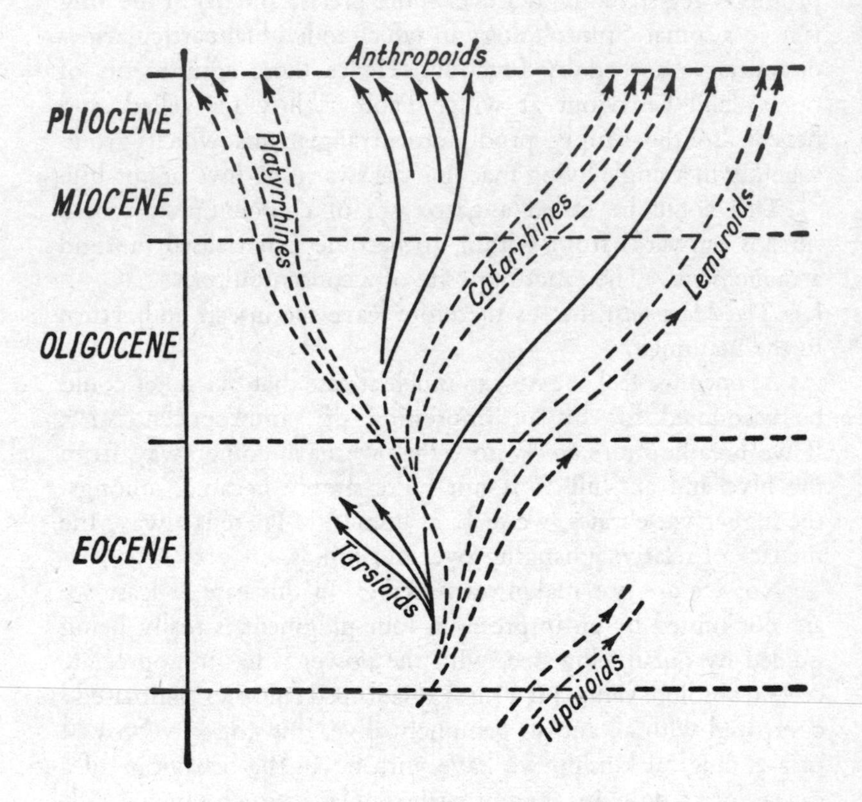


DIAGRAM 3. *The development of the Primates.*

to float, a glimmer of personality. And it is in that direction that the possibilities presently crop up, interminate and interminable, straight ahead.

But from what species was it that the leap forward towards promised horizons was to take place ?

Let us take a closer look at the great horde of Pliocene animals —those limbs developed to the last degree of simplicity and perfection, those forests of antlers on the heads of stags, of lyre-shaped horns on the starred or striped foreheads of antelopes, those heavy tusks on the snouts of the proboscidians, those canines and incisors of the great carnivores . . . Surely such luxuriance, such achievement, must precisely serve to condemn the future of these magnificent creatures, marking them for an early death, writing them off—despite their psychic vitality—as forms that have got into a morphological dead end. All this may seem rather more like an end than a beginning.

This is doubtless so. But besides the Polycladida, the Strepsicerata, the elephants, the sabre-toothed tigers, and so many others, *there are the primates.*

*c. The Primates.* So far I have only mentioned the primates once or twice in passing. I have not yet allotted a place to these near neighbours of ours on the tree of life. The omission was deliberate. At the point I had then reached, their importance had not yet come to light ; they could not have been understood. Now that we have perceived the secret spring that moves zoological evolution, it is different. Their hour has come, and we see how they can and should make their entrance at that fateful moment towards the end of the Tertiary era.

On the whole, like all other animal groups, the primates appear morphologically as a series of overlapping verticils or ramifications, and, as usual, the terminals are sharply defined, the stems blurred (Diag. 3). At the top we have on either side the two great branches of the monkeys proper : the Catarrhines, true monkeys of the old world with 32 teeth ; and the Platyrrhines of South America, with flattened nose and 36 teeth. Below, the lemurs, generally with an elongated snout and often proclivous incisors. Right at the bottom these two-tiered verticils seem to break off at the beginning of the Tertiary era from an 'insectivorous' ramification, the Tupaioids, of which they seem to represent a single radiation in a state of florescence. Nor is that all. At the heart of each of these two verticils we can distinguish a central sub-verticil of particularly 'cephalised' forms. On the lemurian side, the Tarsioids, tiny jumping animals with a round, bulging cranium and huge eyes, whose sole living representative, the tarsier of Malaya, reminds us bizarrely of a little man. On the side of the Catarrhines we are all familiar with anthropoids (the gorilla, the chimpanzee, the orang-outang and the gibbon), tailless monkeys, the biggest and most alert of all monkeys.

The lemurs and the tarsiers were the first to reach their prime—towards the end of the Eocene age. As for the anthropoids, we find them in Africa from the Oligocene onwards. But they certainly did not reach their maximum diversity and size until the end of the Pliocene. Then we find them in both Africa and India, i.e. always in tropical or sub-tropical zones. We should keep in mind this date and this mode of distribution for they contain a lesson.

With that, we have placed the primates outwardly—both in duration and in their external form. We should now penetrate to the *within* of things and try to understand in what respect these animals differ from the others, seen from inside.

What at once catches the anatomist's attention when he looks at monkeys (particularly the higher ones) is the astonishingly slight degree of differentiation in their bones. The cranial capacity is relatively much bigger than in any other mammal, but what are we to say of the rest ? An isolated molar belonging to a Dryopithecus or a chimpanzee could readily be confused with the tooth of an Eocene omnivore such as the Condylarths. Then the limbs—with their radiations still intact these exhibit the same plan and proportions that they had in the first tetrapods of the Palaeozoic era. In the course of the Tertiary era, the ungulates radically transformed the adjustment of their feet ; the carnivores reduced and sharpened their teeth ; the Cetacea streamlined themselves like fish ; the Proboscidea greatly complicated their incisors and their molars. Meanwhile the primates on their side had kept their ulna intact and also their fibula ; they jealously hung on to their five fingers ; they remained typically tritubercular. Are we to consider them therefore the conservatives among mammals, the most conservative of all ?

No ; but they have shown themselves to be the most wary.

In itself, at its best, the differentiation of an organ is an immediate factor of superiority. But, because it is irreversible, it also imprisons the animal that undergoes it in a restricted path at the end of which, under the pressure of orthogenesis, it runs the risk of ending up either in monstrosity or in fragility. Specialisation paralyses, ultra-specialisation kills. Palaeontology is littered with such catastrophes. Because, right up to the Pliocene period, the primates remained the most 'primitive' of the mammals as regards their limbs, they remained also the most *free*. And what did they do with that freedom ? They used it to lift themselves through successive upthrusts to the very frontiers of intelligence.

So we have now before us, simultaneously with the true definition of the primate, the answer to the problem which led us to study the primates. 'After the mammals, at the end of the Tertiary era, where will life be able to carry on ?'

What makes the primates so interesting and important to biology is, in the first place, that they represent a phylum of *pure and direct cerebralisation*. In the other mammals too, no doubt, the nervous system and instinct gradually develop. But in them the internal travail was distracted, limited and finally arrested by accessory differentiations. *Pari Passu* with their psychical development, horse, stag and tiger became, like the insect, to some extent prisoners of the instruments of their swift-moving or predatory ways. *For that is what their limbs and teeth had become*. In the case of the primates, on the other hand, evolution went straight to work on the brain, neglecting everything else, which accordingly remained malleable. That is why they are at the head of the upward and onward march towards greater consciousness. *In this singular and privileged case, the particular orthogenesis of the phylum happened to coincide exactly with the principal orthogenesis of life itself :* following Osborn's term which I shall borrow while changing its sense, it is ' aristogenesis ' —and thus unlimited.

Hence this first conclusion that if the mammals form a dominant branch, *the* dominant branch of the tree of life, the primates (i.e. the cerebro-manuals) are its leading shoot, and the anthropoids are the bud in which this shoot ends up.

Thenceforward, it may be added, it is easy to decide where to look in all the biosphere to see signs of what is to be expected. We already knew that everywhere the active phyletic lines grow warm with consciousness towards the summit. But in one well-marked region at the heart of the mammals, where the most powerful brains ever made by nature are to be found, they become red hot. And right at the heart of that glow burns a point of incandescence.

We must not lose sight of that line crimsoned by the dawn. After thousands of years rising below the horizon, a flame bursts forth at a strictly localised point.

Thought is born.

# BOOK THREE

# THOUGHT

CHAPTER ONE

# THE BIRTH OF THOUGHT

## *Preliminary Remark* : The Human Paradox

FROM A purely positivist point of view man is the most mysterious and disconcerting of all the objects met with by science. In fact we may as well admit that science has not yet found a place for him in its representations of the universe. Physics has succeeded in provisionally circumscribing the world of the atom. Biology has been able to impose some sort of order on the constructions of life. Supported both by physics and biology, anthropology in its turn does its best to explain the structure of the human body and some of its physiological mechanisms. But when all these features are put together, the portrait manifestly falls short of the reality. Man, as science is able to reconstruct him today, is an animal like the others—so little separable anatomically from the anthropoids that the modern classifications made by zoologists return to the position of Linnaeus and include him with them in the same super-family, the hominidae. Yet, to judge by the biological results of his advent, is he not in reality something altogether different ?

Morphologically the leap was extremely slight, yet it was the concomitant of an incredible commotion among the spheres of life—there lies the whole human paradox ; and there, in the same breath, is the evidence that science, in its present-day reconstructions of the world, neglects an essential factor, or rather, an entire dimension of the universe.

In conformity with the general hypothesis which throughout

this book has been leading us towards a coherent and expressive interpretation of the earth as it appears today, I want to show now, in this part devoted to thought, that, to give man his *natural* position in the world of experience, it is necessary and sufficient to consider the *within* as well as the *without* of things. This method has already enabled us to appreciate the grandeur and the direction of the movement of life ; and this method will serve once again to reconcile in our eyes the insignificance and the supreme importance of the phenomenon of man in an order that harmoniously re-descends on life and matter.

Between the last strata of the Pliocene period, in which man is absent, and the next, in which the geologist is dumbfounded to find the first chipped flints, what has happened ? And what is the true measure of this leap ?

It is our task to divine and to measure the answers to these questions before we follow step by step the march of mankind right down to the decisive stage in which it is involved today.

## 1. THE THRESHOLD OF REFLECTION

### A. *The Threshold of the Element : the Hominisation*[1] *of the Individual*

*a. Nature.* Biologists are not yet agreed on whether or not there is a direction (still less a definite axis )of evolution ; nor is there any greater agreement among psychologists, and for a connected reason, as to whether the human psychism differs specifically (by ' nature ') from that of man's predecessors or not. As a matter of fact the majority of ' scientists ' would tend to contest the validity of such a breach of continuity. So much has been said, and is still said, about the intelligence of animals.

If we wish to settle this question of the ' superiority ' of man over the animals (and it is every bit as necessary to settle it for the sake of the ethics of life as well as for pure knowledge) I can

[1] [French : *hominisation*—a word coined by the author.]

only see one way of doing so—to brush resolutely aside all those secondary and equivocal manifestations of inner activity in human behaviour, making straight for the central phenomenon, *reflection*.

From our experimental point of view, reflection is, as the word indicates, the power acquired by a consciousness to turn in upon itself, to take possession of itself *as of an object* endowed with its own particular consistence and value : no longer merely to know, but to know oneself ; no longer merely to know, but to know that one knows.[1] By this individualisation of himself in the depths of himself, the living element, which heretofore had been spread out and divided over a diffuse circle of perceptions and activites, was constituted for the first time as a *centre* in the form of a point at which all the impressions and experiences knit themselves together and fuse into a unity that is conscious of its own organisation.

Now the consequences of such a transformation are immense, visible as clearly in nature as any of the facts recorded by physics or astronomy. The being who is the object of his own reflection, in consequence of that very doubling back upon himself, becomes in a flash able to raise himself into a new sphere. In reality, another world is born. Abstraction, logic, reasoned choice and inventions, mathematics, art, calculation of space and time, anxieties and dreams of love—all these activities of *inner life* are nothing else than the effervescence of the newly-formed centre as it explodes onto itself.

This said, I have a question to ask. If, as follows from the foregoing, it is the fact of being 'reflective' which constitutes the strictly 'intelligent' being, can we seriously doubt that intelligence is the evolutionary lot proper to man and to man *only* ? If not, can we, under the influence of some false modesty, hesitate to admit that man's possession of it constitutes a radical advance on all forms of life that have gone before him ? Admittedly the animal knows. *But it cannot know that it knows* : that is quite certain. If it could, it would long ago have multiplied

[1] [*Non plus seulement connaître, mais se connaître; non plus seulement savoir, mais savoir que l'on sait.*]

its inventions and developed a system of internal constructions that could not have escaped our observation. In consequence it is denied access to a whole domain of reality in which we can move freely. We are separated by a chasm—or a threshold—which it cannot cross. Because we are reflective we are not only different but quite other. It is not merely a matter of change of degree, but of a change of nature, resulting from a change of state.

So we find ourselves confronted with exactly what we expected at the end of the chapter we called *Demeter*. Life, being an ascent of consciousness, could not continue to advance indefinitely along its line without transforming itself in depth. It had, we said, like all growing realities in the world, to become different so as to remain itself. Here, in the accession to the power of reflection, emerges (more clearly recognisable than in the obscure primordial psychism of the first cells) the particular and critical form of transformation in which this surcreation or rebirth consisted for it. And at the same moment we find the whole curve of biogenesis reappearing summed up and clarified in this singular point.

*b. Theoretical Mechanism.* All along, naturalists and philosophers have held opinions of the utmost divergence concerning the 'psychical' make-up of animals. For the early Schoolmen instinct was a sort of sub-intelligence, homogeneous and fixed, marking one of the ontological and logical stages by which being grades downwards from pure spirit to pure materiality. For the Cartesian only thought existed: so the animal, devoid of any *within*, was a mere automaton. For most modern biologists, as I have said already, there is no sharp line to be drawn between instinct and thought, neither being very much more than a sort of luminous halo enveloping the play—the only essential thing—of the determinisms of matter.

In each of these varying opinions there is an element of truth, but also a cause of error which becomes apparent when, following the point of view put forward in these pages, we make up our minds to recognise (1) that instinct, far from being an epiphenomenon, translates through its different expressions the very phenomenon of life, and (2) that it consequently represents a *variable* dimension.

What exactly happens if we look at nature from this angle ?

Firstly we realise better in our minds the fact and the reason for the *diversity* of animal behaviour. From the moment we regard evolution as primarily psychical transformation, we see there is not *one* instinct in nature, but a multitude of forms of instincts each corresponding to a particular solution of the problem of life. The 'psychical' make-up of an insect is not and cannot be that of a vertebrate ; nor can the instinct of a squirrel be that of a cat or an elephant : this in virtue of the position of each on the tree of life.

By the fact itself, in this variety, we begin to see legitimately a relief stand out and a gradation formed. If instinct is a variable dimension, *the* instincts will not only be different ; they constitute beneath their complexity, a growing system. They form as a whole a kind of fan-like structure in which the higher terms on each nervure are recognised each time by a greater range of choice and depending on a better defined centre of co-ordination and consciousness. And that is the very thing we see. The mind (or psyche) of a dog, despite all that may be said to the contrary, is positively superior to that of a mole or a fish.[1]

This being said, and I am merely presenting in a different light what has already been revealed in our study of life, the upholders of the spiritual explanation have no need to be disconcerted when they see, or are obliged to see, in the higher animals (particularly in the great apes) ways and reactions which strangely recall those of which they make use to define the

[1] From this point of view it could be said that every form of instinct tends in its own way to become 'intelligence' ; but it is only in the human line that (for extrinsic or intrinsic reasons) the operation has been successful all the way. Having reached the stage of reflection, man would thus represent a single one of the innumerable modalities of consciousness tried out by life in the animal world. In all those other psychological worlds it is very difficult for us to enter, not only because in them knowledge is more confused, but because they work differently from ours.

nature and prove the presence in man of 'a reasonable soul'. If the story of life is no more than a movement of consciousness veiled by morphology, it is inevitable that, towards the summit of the series, in the proximity of man, the 'psychical' make-ups seem to reach the *borders of intelligence*. And that is exactly what happens.

Hence light is thrown on the 'human paradox' itself. We are disturbed to notice how little 'anthropos' differs anatomically from the other anthropoids, despite his incontestable mental pre-eminence in certain respects—so disturbed that we feel almost ready to abandon the attempt to distinguish them, at least towards their point of origin. But is not this extraordinary resemblance precisely what had to be ?

When water is heated to boiling point under normal pressure, and one goes on heating it, the first thing that follows—without change of temperature—is a tumultuous expansion of freed and vaporised molecules. Or, taking a series of sections from the base towards the summit of a cone, their area decreases constantly ; then suddenly, with another infinitesimal displacement, the surface vanishes leaving us with a *point*. Thus by these remote comparisons we are able to imagine the mechanism involved in the critical threshold of reflection.

By the end of the Tertiary era, the psychical temperature in the cellular world had been rising for more than 500 million years. From branch to branch, from layer to layer, we have seen how nervous systems followed *pari passu* the process of increased complication and concentration. Finally, with the primates, an instrument was fashioned so remarkably supple and rich that the step immediately following could not take place without the whole animal psychism being as it were recast and consolidated on itself. Now this movement did not stop, for there was nothing in the structure of the organism to prevent it advancing. When the anthropoid, so to speak, had been brought 'mentally' to boiling point some further calories were added. Or, when the anthropoid had almost reached the summit of the cone, a final effort took place along the axis. No more was needed for the whole inner equilibrium to be upset. What was previously only a centred surface became a centre. By a tiny 'tangential' increase, the 'radial' was turned back on itself and so to speak took an infinite leap forward. Outwardly, almost nothing in the organs had changed. But in depth, a great revolution had taken place : consciousness was now leaping and boiling in a space of super-sensory relationships and representations ; and simultaneously consciousness was capable of perceiving itself in the concentrated simplicity of its faculties. And all this happened for the first time.[1]

Those who adopt the spiritual explanation are right when they defend so vehemently a certain transcendence of man over the rest of nature. But neither are the materialists wrong when they maintain that man is just one further term in a series of animal forms. Here, as in so many cases, the two antithetical kinds of evidences are resolved in a movement—provided that in this movement we emphasise the highly natural phenomenon of the 'change of state'. From the cell to the thinking animal, as from the atom to the cell, a single process (a psychical kindling or concentration) goes on without interruption and always in the same direction. But by virtue of this permanence in the operation, it is inevitable from the point of view of physics that certain leaps suddenly transform the subject of the operation.

*c. Realisation.* Discontinuity in continuity : that is how, in the theory of its mechanism, the birth of thought, like that of life, presents itself and defines itself.

[1] Need I repeat that I confine myself here to the phenomena, i.e. to the experimental relations between consciousness and complexity, without prejudging the deeper causes which govern the whole issue ? In virtue of the limitations imposed on our sensory knowledge by the play of the temporo-spatial series, it is only, it seems, *under the appearances* of a critical point that we can grasp experimentally the 'hominising' (spiritualising) step to reflection. But, with that said, there is nothing to prevent the thinker who adopts a spiritual explanation from positing (for reasons of a higher order and at a later stage of his dialectic), *under the phenomenal veil* of a revolutionary transformation, whatever 'creative' operation or 'special intervention' he likes (see Prefatory Note). Is it not a principle universally accepted by Christian thought in its theological interpretation of reality that for our minds there are different and successive planes of knowledge ?

But how has the mechanism worked in its concrete reality? Had there been a witness to the crisis, what would have been externally visible to him of the metamorphosis?

As I shall be saying later on, when I come to deal with the 'primaeval forms of man', this picture we are so eager to paint will probably, like the origin of life, remain for ever beyond our grasp—and for the same reasons. The most we have to guide us here is the resource of thinking of the awakening of intelligence in the child in the course of ontogeny. Two remarks deserve, however, to be made, the one circumscribing, the other still further deepening, the mystery which veils this singular point from our imagination.

The first is that to culminate in man at the stage of reflection, life must have been preparing a whole group of factors for a long time and simultaneously—though nothing at first sight could have given grounds for supposing that they would be linked together 'providentially'.

It is true that in the end, from the organic point of view, the whole metamorphosis leading to man depends on the question of a better brain. But how was this cerebral perfectioning to be carried out—how could it have worked—if there had not been a whole series of other conditions realised at just the same time? If the creature from which man issued had not been a biped, his hands would not have been free in time to release the jaws from their prehensile function, and the thick band of maxillary muscles which had imprisoned the cranium could not have been relaxed. It is thanks to two-footedness freeing the hands that the brain was able to grow; and thanks to this, too, that the eyes, brought closer together on the diminished face, were able to converge and fix on what the hands held and brought before them—the very gesture which formed the external counterpart of reflection. In itself this marvellous conjunction should not surprise us. Surely the smallest thing formed in the world is always the result of the most formidable coincidence—a knot whose strands have been for all time converging from the four corners of space. Life does not work by following a single thread, nor yet by fits and starts. It pushes forward its whole network at one and the same time. So is the embryo fashioned in the womb that bears it. This we have reason to know, but it is satisfying to us precisely to recognise that man was born under the same maternal law. And we are happy to admit that the birth of intelligence corresponds to a turning in upon itself, not only of the nervous system, but of the whole being. What at first sight disconcerts us, on the other hand, is the need to accept that this step could only be achieved *at one single stroke*.

For that is to be my second remark, a remark I cannot avoid. In the case of human ontogeny we can slur over the question at what moment the new-born child may be said to achieve intelligence and become a thinking being, for we find a continuous series of *states* happening in the same individual from the fertilised ovum to the adult. What does it matter whether there is a hiatus or where it might be ? It is quite different in the case of a phyletic embryogenesis in which each stage or each state is represented by a *different being*, and it is impossible (at any rate within the scope of modern methods of thought) to evade the problem of discontinuity. If the threshold of reflection is really (as its physical nature seems to require, and as we have ourselves admitted) a critical transformation, a mutation from zero to everything, it is impossible for us to imagine an intermediary individual at this precise level. Either this being has not yet reached, or it has already got beyond, this change of state. Look at it as we will, we cannot avoid the alternative—either thought is made unthinkable by a denial of its psychical transcendence over instinct, or we are forced to admit that it appeared *between* two individuals.

The terms of this proposition are disconcerting, but they become less bizarre, and even inoffensive, if we observe that, speaking strictly as scientists, we may suppose that intelligence might (or even must) have been as little visible externally at its phyletic origin as it is today to our eyes in every new-born child at the ontogenetical stage: in which case every tangible subject of debate between the observer and the theorist disappears.

To say nothing of the fact (see the second form of the 'ungraspable' in the footnote on p. 186) that any sort of scientific discussion today on the outward and visible signs of the first emergence of reflection on the earth (even supposing there had been a spectator there to see them) is quite impossible ; because, here if anywhere, we find ourselves in the presence of one of those *beginnings* ('infinitely small quantities in evolution') automatically and irremediably removed from our range of vision by a thick layer of the past (see Note, p. 122).

Without trying to picture the unimaginable, let us therefore keep hold of one idea—that the access to thought represents a threshold which had to be crossed at a single stride ; a 'trans-experimental' interval about which scientifically we can say nothing, but beyond which we find ourselves transported onto an entirely new biological plane.

*d. Prolongation.* It is only at this point that we can fully see the nature of the transit to reflection. In the first place it involved a change of state ; then, by this very fact, the beginning of another kind of life—precisely that interior life of which I have spoken above. A moment ago we compared the simplicity of the thinking mind with that of a geometrical point. It would have been better to speak of a line or an axis. Where intelligence is concerned, 'to be posited' does not mean 'to be achieved'. As soon as a child is born, it must breathe or it will die. Similarly the reflective psychic centre, once turned in upon itself, can only subsist by means of a double movement which is in reality one and the same. It centres itself further on itself by penetration into a new space, and at the same time it centres the rest of the world around itself by the establishment of an ever more coherent and better organised perspective in the realities which surround it. We are not dealing with an immutably fixed focus but with a vortex which grows deeper as it sucks up the fluid at the heart of which it was born. The ego only persists by becoming ever more itself, in the measure in which it makes everything else itself. *So man becomes a person in and through personalisation.*

Obviously by the effect of such a transformation the entire

structure of life is modified. Up to this point the animated element was so narrowly subject to the phylum that its own individuality could be regarded as accessory and sacrificed. It received, maintained, acquired if possible, reproduced and transmitted. And so on ceaselessly and indefinitely. Caught up in the chain of succeeding generations, the animal seemed to lack the right to live ; it appeared to have no value for itself. It was a fugitive foothold for a process which passed over it and ignored it. Life, once again, was more real than living things.

With the advent of the power of reflection (an essentially elemental property, at any rate to begin with) everything is changed, and we now perceive that under the more striking reality of the collective transformations a secret progress has been going on parallel to individualisation. The more highly each phylum became charged with psychism, the more it tended to 'granulate'. The animal grew in value in relation to the species. Finally at the level of man the phenomenon gathers new power and takes definitive shape. With the 'person', endowed by 'personalisation' with an indefinite power of elemental evolution, the branch ceases to bear, as an anonymous whole, the exclusive promises for the future. The cell has become 'someone'. After the grain of matter, the grain of life ; and now at last we see constituted the *grain of thought.*

Does that mean that the phylum loses its function from this moment and vanishes in thin air, like those animals who lose their identity in a veritable dust of spores which they give birth to in dying ? Above the point of reflection, does the whole interest of evolution shift, passing from life into a plurality of isolated living beings ?

Nothing of the sort. Only, from this crucial date the global spurt, without slackening in the slightest, has acquired another degree, another order of complexity. The phylum does not break like a fragile jet just because henceforward it is fraught with thinking centres ; it does not crumble into its elementary psychisms. On the contrary it is reinforced by an inner lining, an additional framework. Until now it was enough to consider in nature a simple vibration on a wide front, the ascent of individual centres of consciousness. What we now have to do is to define and regulate harmoniously an ascent of consciousnesses (a much more delicate phenomenon). We are dealing with a progress made up of other progresses as lasting as itself ; a movement of movements.

Let us try to lift our minds high enough to dominate the problem. For that, let us forget for a moment the particular destiny of the spiritual elements engaged in the general transformation. It is, in point of fact, only by following the ascension and spread of the whole in its main lines that we are able, after a long detour, to determine the part reserved for individual hopes in the total success.

We thus reach the personalisation of the individual by the 'hominisation' of the whole group.

### B. *The Threshold of the Phylum : the Hominisation of the Species*

Thus, through this leap of intelligence, whose nature and mechanism we have been analysing in the thinking particle, life continues in some way to spread as though nothing had happened. According to all appearances, propagation, multiplication and ramification went on in man, as in other animals, after the threshold of thought, as busily as before. Nothing, one might think, had altered in the current. But the water in it was no longer the same. Like a river enriched by contact with an alluvial plain, the vital flux, as it crossed the stages of reflection, was charged with new principles, and as a result manifested new activities. From now onwards it was not merely animated grains which the pressure of evolution pumped up the living stem, but grains of thought. What was to happen under this influence to the colour or the shape of the leaves, the flowers, the fruit ?

I would be anticipating later developments of our argument if I gave a detailed and considered answer to this question now.

But it would be as well to indicate at once three particularities which manifest themselves in any and every operation or production of the species from the moment the threshold of thought is crossed. One concerns the composition of new branches, another the general direction of their growth, the third their relations to and differences from—taken as a whole—what had flourished earlier on the tree of life.

*a. The composition of the human branches.* Whatever idea we have about the inner mechanism of evolution, there is no denying that each zoological group is enclosed in a certain psychological envelope. We have already said that each type of insect, bird or mammal has its own instincts. So far no attempt has been made to link together systematically the two elements, namely the somatic and psychic, of the species. There are naturalists who describe and classify shapes, and others who specialise in the study of behaviour. In fact, below man, purely morphological criteria provide a perfectly adequate framework for studying the distribution of species. But from the advent of man difficulties appear. We cannot fail to be aware of the extreme confusion which prevails concerning the significance and the distribution of the extremely varied groups into which mankind divides up under our very eyes—races, nations, states, countries, cultures, etc. In these diverse and constantly shifting categories, people as a rule only care to see heterogeneous units—some natural (race), others artificial (nations)—overlapping irregularly on different planes.

It is an unpleasing and unnecessary irregularity, and one which vanishes as soon as we give its proper place to the *within* as well as to the *without* of things.

Indeed, from this more comprehensive point of view, the composition of the human group with its branches, however confused it may appear, can be reduced nevertheless to the general rules of biology. But, by the exaggeration of a variable that had remained negligible in the animals, it simply brings out the dual nature of those rules or even, on the contrary—if what is somatic is woven by the psyche—their fundamental unity. This is not an exception but a generalisation. It is impossible to remain long in doubt : in the world become human it is always the zoological ramification which, in spite of all appearances and all complexities, pushes onwards and operates according to the same mechanism as before. Only, as a result of the quantity of inner energy liberated by reflection, the operation then tends to emerge from the material organs so as to formulate itself *also* or even *above all* in the mind. What is spontaneously psychical is no longer merely an aura round the 'soma'. It becomes an appreciable part, or even a principal part, of the phenomenon. And because variations of soul are much richer and more subtle than the often imperceptible organic changes which accompany them, it is obvious that the mere inspection of bones or integuments will not suffice to explain or to catalogue the progresses of the total zoological differentiation. That is how things stand. And the remedy faces us no less clearly. To unravel the structure of a thinking phylum, anatomy by itself is not enough : it must be backed up by psychology.

This is a laborious complication of course, since it becomes clear that no satisfactory classification of the human 'genus' will be forthcoming, save through the combined play of two partially independent variable. But it is a fruitful complication, for two reasons.

On the one hand, at the price of this difficulty, order and homogeneity—that is to say, truth—come back into our perspectives of life extended to include man ; and, because we realise correlatively the organic value of every social construction, we feel already more inclined to treat it as a subject of science, hence to respect it.

On the other hand, from the very fact that the fibres of the human phylum appear surrounded by their psychic sheath, we can begin to understand the extraordinary power of agglutination and coalescence that they show. Which brings us at the same time on the track of the fundamental discovery with which our study of the phenomenon of man is to culminate—the convergence of the spirit.

*b. The General Direction of Growth.* So long as our perspectives of the psychic nature of zoological evolution were based only on the examination of animal lines and their nervous systems, the direction of that evolution remained perforce as vague for our knowledge as the soul itself of those distant relations of ours. Consciousness rises through living beings : that was about all we were able to say. But from the moment the threshold of thought is crossed its progress becomes easier to unravel ; for life has not only reached the rung on which we ourselves stand, but begins to overflow freely by its free activity beyond the boundary within which it had been confined by the exigences of physiology. The message is more clearly written, and we are better able to follow it, because we recognise ourselves in it. Earlier, when we were discussing the tree of life, we noticed as a fundamental character that brains grew bigger and became more differentiated along each zoological stem. To define the extension and the counterpart of this law (after the transit to reflection) it will henceforth be sufficient to say : 'Following each anthropological line, it is the human element that seeks itself and grows.'

A moment ago I referred to the unparalleled complexity of the human group—all those races, those nations, those states whose entanglements defy the resourcefulness of anatomists and ethnologists alike. There are so many rays in that spectrum that we despair of analysing them. Let us try instead to perceive what this multiplicity represents when viewed as a whole. If we do this we will see that its disturbing aggregation is nothing but a multitude of sequins all sending back to each other by reflection the same light. We find hundreds or thousands of facets, each expressing at a different angle a reality which seeks itself among a world of groping forms. We are not astonished (because it happens to *us*) to see in each person around us the spark of reflection developing year by year. We are all conscious, too, at all events vaguely, that *something* in our atmosphere is changing with the course of history. If we add these two pieces of evidence together (and rectify certain exaggerated views on the purely 'germinal' and passive nature of heredity), how is it that we are not more sensitive to the presence of something greater than ourselves moving forward within us and in our midst ?

Up to the level of thought a question could still be asked of the science of nature—the question about the evolutionary value and transmission of acquired characters. As we know, the biologist tended, and still tends, to be sceptical and evasive ; and perhaps he is right, as regards the fixed zones of the body he likes to confine himself to. But what happens if we give the psyche its legitimate place in the integrity of living organisms ? Immediately, over the alleged independence of the phyletic 'germ-plasm', the individual activity of the 'soma' reclaims its rights. In the insects, for example, or the beaver, we see in the most blatant way the existence of hereditarily-formed or even fixed instincts underlying the play of animal spontaneities. From reflection onwards, the reality of this mechanism becomes not only manifest but preponderant. Under the free and ingenious effort of successive intelligences, *something* (even in the absence of any measurable variation of brain or cranium) irreversibly accumulates, according to all the evidence, and is transmitted, at least collectively by means of education, down the course of ages. The point here is that this 'something'—construction of matter or construction of beauty, systems of thought or systems of action—ends up always by translating itself into an augmentation of consciousness, and consciousness in its turn, as we now know, is nothing less than the substance and heart of life in process of evolution.

What can this mean except that, over and above this particular phenomenon—the individual accession to reflection—science has grounds for recognising another phenomenon of a reflective nature co-extensive with the whole of mankind. Here as elsewhere in the universe, the whole shows itself to be greater than the simple sum of the elements of which it is formed. The human individual does not exhaust in himself the vital potentialities of his race. But following each strand known to anthropology and sociology, we meet with a stream whereby a continuing and transmissible tradition of reflection is established and allowed to increase. So from individual men there springs the human reality ; from human phylogenesis, the human stem.[1]

c. *Connections and Differences.* That seen and accepted, under what form should we expect the human stem to rise up ? Will it, because it is a thinking stem, sever the fibres which attach it to the past—and, at the summit of the vertebrate branch, will it develop from new elements and according to a new plan, like some neoplasm ? To imagine such a rupture would be, once again, to misjudge and underestimate our own ' dimension ' as well as the organic unity of the world and the methods of evolution. In a flower the sepals, petals, stamens and pistil are not leaves and they have probably never been leaves. Yet they possess unmistakably in their attachments and their texture everything that would have resulted in a leaf had they not been formed under a new influence and a new destiny. Similarly, with the human inflorescence, we can see transformed or undergoing transformation the vessels, the disposition, and even the sap of the stalk upon which the inflorescence was born : not only the individual structure of the organs and the interior ramifications of the species, but even the tendencies and behaviour of the ' soul '.

In man, considered as a zoological group, everything is extended simultaneously—sexual attraction, with the laws of reproduction ; the inclination to struggle for survival, with the competitions it involves; the need for nourishment, with the accompanying taste for seizing and devouring ; curiosity, to see, with its delight in investigation ; the attraction of joining others to live in society. Each of these fibres traverses each one of us, coming up from far below and stretching beyond and above us. And each one of them has its story (no less true

[1] [Even if the Lamarckian view of the heritability of acquired characteristics is biologically *vieux jeu*, and decisively refuted, when we reach the human level and have to reckon with history, culture etc., ' transmission ' becomes ' tradition '. See M. Polanyi, *Personal Knowledge* (Kegan Paul, 1958)].

than any other) to tell of the whole course of evolution—evolution of love, evolution of war, evolution of research, evolution of the social sense. But each one, just because it is evolutionary, undergoes a metamorphosis as it crosses the threshold of reflection. Beyond this point it is enriched by new possibilities, new colours, new fertility. It is the same thing, if you like, but it is something quite different also—a figure that has become transformed by a change of space and dimension, discontinuity superimposed upon continuity, mutation upon evolution.

In this supple inflection, in this harmonious recasting which transfigures the whole grouping of vital antecedences, both external and internal, we cannot fail to find precious confirmation of what we had already guessed. When an object begins to grow in one of its accessory parts, it is thrown out of equilibrium and becomes deformed. To remain symmetrical and beautiful a body must be modified simultaneously throughout, in the direction of one of its principal axes. Reflection conserves even while re-shaping all the lines of the phylum on which it settles. There is no fortuitous excrescence of a parasitic energy. Man only progresses by slowly elaborating from age to age the essence and the totality of a universe deposited within him.

To this grand process of sublimation it is fitting to apply with all its force the word *hominisation*. Hominisation can be accepted in the first place as the individual and instantaneous leap from instinct to thought, but it is also, in a wider sense, the progressive phyletic spiritualisation in human civilisation of all the forces contained in the animal world.

Thus we are led—after having considered the element and pictured the species —to contemplate the earth in its totality.

### c. *The Threshold of the Terrestrial Planet : the Noosphere*

When compared to all the living verticils, the human phylum is not like any other. But because the specific orthogenesis of the primates (urging them towards increasing cerebralisation) coincides with the axial orthogenesis of organised matter (urging all living things towards a higher consciousness) man, appearing at the heart of the primates, flourishes on the leading shoot of zoological evolution. It was with this observation that we rounded off our remarks on the state of the Pliocene world.

It is easy to see what privileged value that unique situation will confer upon the transit to reflection.

'The biological change of state terminating in the awakening of thought does not represent merely a critical point that the individual or even the species must pass through. Vaster than that, it affects life itself in its organic totality, and consequently it marks a transformation affecting the state of the entire planet.'

Such is the evidence—born of all the other testimony we have gradually assembled and added together in the course of our nquiry—which imposes itself irresistibly on both our logic and observation.

We have been following the successive stages of the same grand progression from the fluid contours of the early earth. Beneath the pulsations of geo-chemistry, of geo-tectonics and of geo-biology, we have detected one and the same fundamental process, always recognisable—the one which was given material form in the first cells and was continued in the construction of nervous systems. We saw geogenesis promoted to biogenesis, which turned out in the end to be nothing else than psychogenesis.

With and within the crisis of reflection, the next term in the series manifests itself. Psychogenesis has led to man. Now it effaces itself, relieved or absorbed by another and a higher function—the engendering and subsequent development of the mind, in one word *noogenesis*. When for the first time in a living creature instinct perceived itself in its own mirror, the whole world took a pace forward.

As regards the choices and reponsibilities of our activity, the consequences of this discovery are enormous. As regards our understanding of the earth they are decisive.

Geologists have for long agreed in admitting the zonal composition of our planet. We have already spoken of the barysphere, central and metallic, surrounded by the rocky lithosphere that in turn is surrounded by the fluid layers of the hydrosphere and the atmosphere. Since Suess, science has rightly become accustomed to add another to these four concentric layers, the living membrane composed of the fauna and flora of the globe, the biosphere, so often mentioned in these pages, an envelope as definitely universal as the other 'spheres' and even more definitely individualised than them. For, instead of representing a more or less vague grouping, it forms a single piece, of the very tissue of the genetic relations which delineate the tree of life.

The recognition and isolation of a new era in evolution, the era of noogenesis, obliges us to distinguish correlatively a support proportionate to the operation—that is to say, yet another membrane in the majestic assembly of telluric layers. A glow ripples outward from the first spark of conscious reflection. The point of ignition grows larger. The fire spreads in ever widening circles till finally the whole planet is covered with incandescence. Only one interpretation, only one name can be found worthy of this grand phenomenon. Much more coherent and just as extensive as any preceding layer, it is really a new layer, the 'thinking layer', which, since its germination at the end of the Tertiary period, has spread over and above the world of plants and animals. In other words, outside and above the biosphere there is the noosphere.

With that it bursts upon us how utterly warped is every classification of the living world (or, indirectly, every construction of the physical one) in which man only figures logically as a *genus* or a new family. This is an error of perspective which deforms and uncrowns the whole phenomenon of the universe. To give man his true place in nature it is not enough to find one more pigeon-hole in the edifice of our systematisation or even an additional order or branch. With hominisation, in spite of the insignificance of the anatomical leap, we have the beginning of a new age. The earth 'gets a new skin'. Better still, it finds its soul.

Therefore, given its place in reality in proper dimensions, the historic threshold of reflection is much more important than any zoological gap, whether it be the one marking the origin of the tetrapods or even that of the metazoa. Among all the stages successively crossed by evolution, the birth of thought comes directly after, and is the only thing comparable in order of importance to, the condensation of the terrestrial chemism or the advent of life itself.

The paradox of man resolves itself by passing beyond measure. Despite the relief and harmony it brings to things, this perspective is at first sight disconcerting, running counter as it does to the illusion and habits which incline us to measure events by their material face. It also seems to us extravagant because, steeped as we are in what is human like a fish in the sea, we have difficulty in emerging from it in our minds so as to appreciate its specificness and breadth. But let us look round us a little more carefully. This sudden deluge of cerebralisation, this biological invasion of a new animal type which gradually eliminates or subjects all forms of life that are not human, this irresistible tide of fields and factories, this immense and growing edifice of matter and ideas—all these signs that we look at, for days on end—to proclaim that there has been a change on the earth and a change of planetary magnitude.

There can indeed be no doubt that, to an imaginary geologist coming one day far in the future to inspect our fossilised globe, the most astounding of the revolutions undergone by the earth would be that which took place at the beginning of what has so rightly been called the psychozoic era. And even today, to a Martian capable of analysing sidereal radiations psychically no less than physically, the first characteristic of our planet would be, not the blue of the seas or the green of the forests, but the phosphorescence of thought.

The greatest revelation open to science today is to perceive that everything precious, active and progressive originally contained in that cosmic fragment from which our world emerged, is now concentrated in a ' crowning ' noosphere.

And what is so supremely instructive about the origins of this noosphere (if we know how to look) is to see how gradually, by dint of being universally and lengthily prepared, the enormous event of its birth took place.

## 2. THE ORIGINAL FORMS

Man came silently into the world.

For a century or so, the scientific problem of the origin of man has been under discussion, and a swelling team of research workers has been digging feverishly into the past to discover the initial point of hominisation, and yet I cannot find a more expressive formula than this to sum up all our prehistoric knowledge. The more we find of fossil human remains and the better we understand their anatomic features and their succession in geological time, the more evident it becomes, by an unceasing convergence of all signs and proofs, that the human ' species ', however unique the ontological position that reflection gave it, did not, at the moment of its advent, make any sweeping change in nature. Whether we consider the species in its environment, in the morphology of its stem, or in the global structure of its group, we see it emerge phyletically exactly *like any other* species.

Firstly, *in its environment*. As we know from palaeontology, an animal form never comes singly. It is sketched out in the heart of a verticil of neighbouring forms among which it takes shape, so to speak, gropingly. So it is with man. Regarded zoologically, man is today an almost isolated figure in nature. In his cradle he was less isolated. Nowadays there is no more room for doubt. Over a well-defined but immense area, extending from South Africa to Southern China and Malaya, amongst the rocks and the forest, at the end of the Tertiary period, the anthropoids were far more numerous than they are today.

Besides the gorilla, chimpanzee and orang-outang, now thrown back into their last strongholds like the Australian bushmen and the negrillos of our day, there was a whole population of other big primates, some of whom (the African Australopithecus, for instance) seem to have been far more hominoid than any alive today.

Secondly, *in the morphology of its stem.* With the multiplication of 'sister-forms', what indicates to the naturalist the origin of a living stem is a certain convergence of the axis of that stem with that of its neighbours. In the proximity of a knot, the leaves grow closer together. Not only is a species at its birth found bunched with others, but, like them it betrays much more clearly than in adult life its zoological parentage. The farther we follow an animal line back into the past, the more numerous and the more palpable are its 'primitive' features. Here too, man, on the whole, keeps strictly to the habitual phyletic mechanism. All we need is to try to arrange in a descending series Pithecanthropus and Sinanthropus after the Neanderthaloids below present-day man. Palaeontology does not often succeed in tracing so satisfying an alignment.

Thirdly, *in the structure of its group.* However well-defined the characters of a phylum may be, it is never found to be altogether simple, like a pure radiation. On the contrary, as far as we can follow it into the depths of its past, it manifests an internal tendency to cleavage and dispersion. Newly born, or even while *being* born, the species breaks up into varieties or sub-species. This is known to all naturalists. Keeping it in mind, let us take another look at man, man whose pre-history (even the most ancient) proves his congenital aptitude for ramification. Is it possible to deny that in the fan of the anthropoids he isolated himself—in this subject to the laws of all animate matter—as a fan of his own ?

I was not exaggerating in the least. The more deeply science plumbs the past of our humanity, the more clearly does it see that humanity, *as a species,* conforms to the rhythm and the rules that marked each new offshoot on the tree of life before the advent of mankind. Thus we are logically obliged to pursue the subject to its conclusion. Since man as a species is at birth so similar to the other phyla, let us stop being surprised if, as with all living groups, the fragile secrets of his earliest origins give our science the slip; and let us henceforward forbear to force and falsify this natural condition with clumsy questionings.

Man came silently into the world. As a matter of fact he trod so softly that, when we first catch sight of him as revealed by those indestructible stone instruments, we find him sprawling all over the old world from the Cape of Good Hope to Peking. Without doubt he already speaks and lives in groups ; he already makes fire. After all, this is surely what we ought to expect. As we know, each time a new living form rises up before us out of the depths of history, it is always complete and already legion.

Thus *in the eyes of science*, which at long range can only see things in bulk, the ' first man ' is, and can only be, *a crowd*, and his infancy is made up of thousands and thousands of years.[1]

It is inevitable that this situation should be disappointing, leaving our curiosity unsatisfied. For what most interests us is precisely what happened during those first thousands of years. And still more, what marked the first critical moment. Dearly would we love to know what those first parents of ours looked like, the ones that stood just this side of the threshold of reflection. As I have already said, that threshold had to be crossed in a single stride. Imagine the past to have been photographed section by section : at that critical moment of initial hominisation, what should we see when we developed our film ?

If we have understood the limits of enlargement imposed by nature on the instrument which helps us to study the landscape

[1] That is why the problem of monogenism in the strict sense of the word (I do not say monophyletism—see below) seems to *elude* science as such by its very nature. At those depths of time when hominisation took place, the presence and the movements of a unique couple are positively ungraspable, unrevealable to our eyes at no matter what magnification. Accordingly one can say that there is room *in this interval* for anything that a trans-experimental source of knowledge might demand.

of the past, we shall be prepared to forgo the satisfaction of this futile curiosity. No photograph could record upon the human phylum this passage to reflection which so naturally intrigues us, for the simple reason that the phenomenon took place inside that which is *always* lacking in a reconstructed phylum—the peduncle of its original forms.

But if the tangible forms of this peduncle escape us, can we not at any rate guess indirectly at its complexity and initial structure ? On these points palaeanthropology has not yet made up its mind. We could, however, try to form an opinion.[1]

A number of anthropologists, and those not the least eminent, think the peduncle of our race must have been composed of several distinct but related 'bundles'. Just as, on the plane of human intellect, once a certain degree of preparation and tension has been reached, the same idea may come to birth at several points simultaneously, so in the same way, man, according to these authorities, must have started simultaneously in several regions on the 'anthropoid layer' of the Pliocene era, thereby following the general mechanism of all life. This is not properly speaking 'polyphyletism', because the different points of germination are located on the same zoological stem, but it is an extensive mutation of the whole stem itself. The idea involves 'hologenesis' and therefore polycentricity. What we get is a whole series of points of hominisation scattered along a subtropical zone of the earth, and hence several human stems becoming genetically merged somewhere *beneath* the threshold of reflection ; not a 'focus' but a 'front' of evolution.

Though not disputing the value and the scientific probabilities of this perspective, I feel myself personally attracted to a slightly

[1] Some idea of how the transit to man was effected zoologically is perhaps suggested by the case of Australopithecus mentioned above. In this South African family of Pliocene anthropomorphs (evidently a group in a state of active mutation) in which a whole series of hominoid characters overlay a basis still clearly simian, we can see an image perhaps, or call it a faint echo, of what was taking place at about the same period even not far from there, in another anthropoid group, in this case culminating in genuine hominisation.

different hypothesis. I have already stressed several times that curious peculiarity shown by zoological branches of bearing fixed on them, like essential characters, certain traits whose origin is plainly peculiar and accidental—such as the tritubercular teeth and seven cervical vertebrae of the higher mammals, the four-footedness of the walking vertebrates, the rotatory power *in one particular direction* of organic substances. Precisely because these traits are secondary and accidental, their universal occurrence in groups, sometimes vast, can only be properly explained by assuming these groups to derive from a highly particularised and therefore extremely localised verticil. We would thus perhaps find no more than a single radiation in a verticil to support originally a layer or even a branch or even the whole of life. Or, if some convergence has played a part, it can only have been amongst closely-related fibres.

In the light of these considerations, and particularly when dealing with a group as homogeneous and specialised as the one under discussion, I feel inclined to minimise the effects of parallelism in the initial formation of the human branch. On the verticil of the higher primates, this branch did not, in my opinion, glean its fibres here and there, one by one, from the whole range offered : but, even more closely than any other species, this branch, I am convinced, represents the thickening and successful development of one solitary stem among all—this stem being, moreover, the most central of the collection because the most vital and, except for the brain, the least specialised. If that is right, all human lines join up genetically, but at the bottom, at the very point of reflection.[1]

And now, if we do assume the strictly unique existence of such a peduncle at the origin of man, what more (still keeping to the plane of pure phenomena) can we say about its length and probable thickness ? Should we, like Osborn, locate its

[1] Which amounts to saying that if the science of man can say nothing directly for or against monogenism (a single initial couple—see note p. 186) it can on the other hand come out decisively, it seems, in favour of monophyletism (a single phylum).

separation very low down, in the Eocene or Oligocene period in a ramification of pre-anthropoid forms ? Or should we, like W. K. Gregory, regard it as a branching off from the anthropoid verticil as late as the Pliocene age ?

Another question, always on the same subject and still maintaining a strictly 'phenomenal' attitude : what *minimum* diameter should we ascribe as biologically possible to this stem (whether it is deep or not) if we consider it at its initial point of hominisation ? For it to be able to 'mutate', resist and live, what is the minimum number of individuals (in order of size) that must have undergone simultaneously the metamorphosis of reflection ? However monophyletic one supposes it to be, surely a species is always sketched out like a diffuse current in a river—by mass effects ? Or, on the contrary, should we rather view it as propagating itself like crystallisation beginning with a few parts —by effect of unities ? In our minds the two symbols (each partly true perhaps) still conflict and have their respective advantages and attractions. We must have the patience to wait until their synthesis is established.

Let us wait. And to encourage our patience let us recall the two following points.

The first is that on every hypothesis, however solitary his advent, man emerged from a general groping of the world. He was born a direct lineal descendant from a total effort of life, so that the species has an axial value and a pre-eminent dignity. At bottom, to satisfy our intelligence and the requirements of our conduct, we have no need to know more than this.

The second point is that, fascinating as the problem of our origin is, its solution even in detail would not solve the problem of man. We have every reason to regard the discovery of fossil men as one of the most illuminating and critical lines of modern research. We must not, however, on that account, entertain any illusions concerning the limits in all its domains of that form of analysis that we call embryogenesis. If in its structure the embryo of each thing is fragile, fleeting and hence, in the past, practically ungraspable, how much more is it ambiguous and undecipherable in its lineaments ? It is not in their germinal state that beings manifest themselves but in their florescence. Taken at the source, the greatest rivers are no more than narrow streams.

To grasp the truly cosmic scale of the phenomenon of man, we had to follow its roots through life, back to when the earth first folded in on itself. But if we want to understand the specific nature of man and divine his secret, we have no other method than to observe what reflection has already provided and what it announces *ahead.*

CHAPTER TWO

# THE DEPLOYMENT OF THE NOOSPHERE

---

IN ORDER to multiply the contacts necessary for its gropings and to be able to store up the multifarious variety of its riches, life is obliged to move forward in terms of deep masses. And when therefore its course emerges from the gorges in which a new mutation has so to speak strangled it, the narrower the channel from which it emerges and the vaster the surface it has to cover with its flow, the more it needs to re-group itself in multitude.

Our picture is of mankind labouring under the impulsion of an obscure instinct, so as to break out through its narrow point of emergence and submerge the earth ; of thought becoming number so as to conquer all habitable space, taking precedence over all other forms of life ; of mind, in other words, deploying and convoluting the layers of the noosphere. This effort at multiplication and organic expansion is, for him who can see, the summing up and final expression of human pre-history and history, from the earliest beginnings down to the present day.

We will now try, in a few bold strokes, to map out the phases or successive waves of this invasion (diag. 4).

## 1. THE RAMIFYING PHASE OF THE PRE-HOMINIDS

Towards the very end of the lower Pleistocene period, a vast upward movement, a positive jolt, seems to have affected the

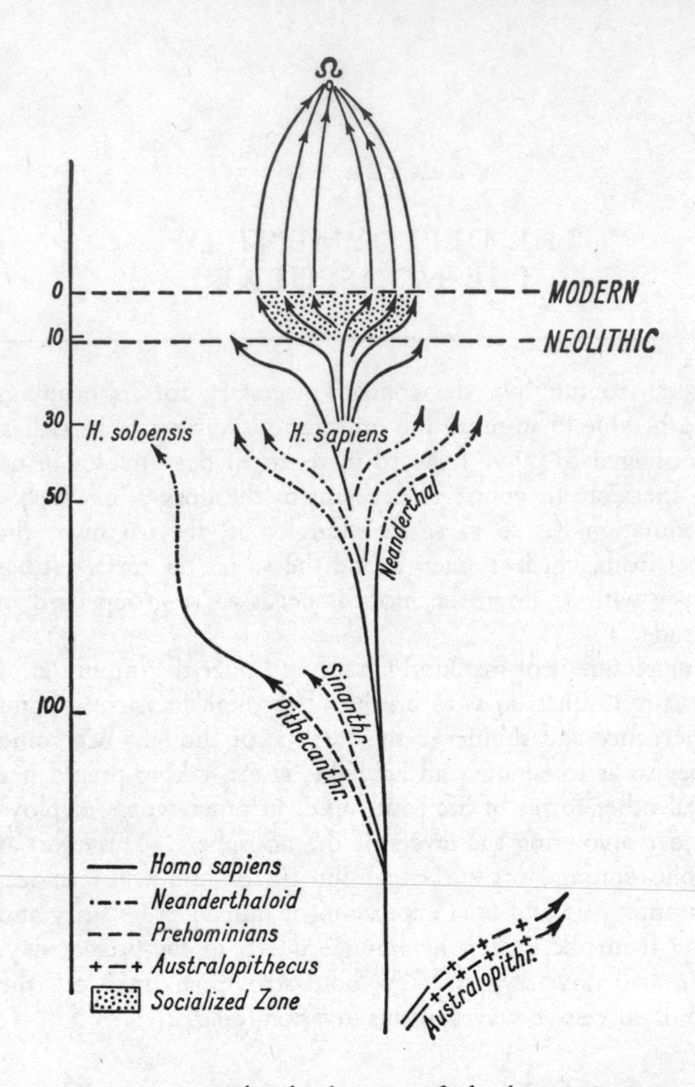


DIAGRAM 4. *The development of the human Layer. The figures on the left indicate thousands of years. They are a minimum estimate and should probably be at least doubled. The hypothetical zone of convergence on the point Omega is obviously not to scale. By analogy with other living layers, its duration should certainly run into thousands of years.*

continental masses of the old world from the Atlantic to the Pacific.[1] Almost everywhere, at this period, we find the land being drained, ravines being carved, and thick layers of alluvium spreading over the plains. Before this great upheaval we can establish no certain trace of man anywhere. Yet it was barely over when we find chipped stone mixed with the gravels on almost all the raised lands of Africa, Western Europe and Southern Asia.

Man of the Lower Quaternary period, the contemporary and the author of these earliest tools is only known to us in two fossil remains. We know them well, however—the *Pithecanthropus* of Java, long represented only by a simple skull, but now by much more satisfactory specimens recently discovered ; and the *Sinanthropus* of China, numerous specimens of which have been found in the last ten years. These two beings are so closely related that the nature of each would have remained obscure if we had not had the good fortune to be able to compare them.[2]

What can we learn from these venerable relics which are at least some one or two hundred thousand years old ?

To begin with, anthropologists are now in agreement on one point : both *Pithecanthropus* and *Sinanthropus* are already definitely hominid *in their anatomy*. If we arrange their skulls in series between those of the great apes on the one hand and modern man on the other, we are at once struck by the wide morphological breach, the void, apparent between them and the anthropoids, while on the human side they seem to fall naturally into the same cast. We find a relatively short face and a relatively spacious cranium. In Trinil man the cerebral capacity hardly descends below 800 c.c. while with Peking man in the biggest

[1] At the end of the Villafranchian age, to be more exact. [By a decision of the International Geological Congress (1948), the Villafranchian is now included in the Pleistocene.]

[2] To avoid complicating the story, I will say nothing here of Heidelberg man. However ancient and remarkable his jaw, we do not know enough about him to determine his real anthropological position.

males it reaches 1100.[1] We find a lower jaw essentially constructed on human lines towards the symphysis, and lastly and most important of all, we find erect biped posture leaving the fore limbs free. With all these signs it is quite obvious that we are on the human side of the line.

However hominid the *Pithecanthropus* and *Sinanthropus* were, judged by their physiognomy they were certainly strange creatures such as have long ago vanished from the earth. Elongated skull, markedly compressed behind enormous orbits; flattened cranium whose transverse section, instead of being ovoid or pentagonal, as with us, forms an arch widely open at the level of the ears; strongly ossified skull whose brain-box does not project backwards but is surrounded posteriorly by a thick occipital roll; a prognathous skull whose dental arches project far forward above a symphysis which not only lacks a chin but is receding; and finally, highly marked sexual dimorphism, the females being small with rather slender jaws and teeth, the males robust with strong molars and canines. These various characters, in no way teratological, but expressing a well-established, well-balanced architecture, seem to suggest, anatomically, a downward convergence towards the 'simian' world.

All things considered, the scientist can affirm without further hesitation that, thanks to the double discovery of Trinil man and Peking man, we recognise within mankind a further morphological rung, a further evolutionary stage and a further zoological verticil.

They are a morphological rung because on the line separating, for instance, a white man from a chimpanzee, we must place them, by the form of their skull, almost exactly half-way;

They are an evolutionary stage because, whether they have or have not left any direct descendants in the contemporary world, they probably represent a type through which modern man must once have passed in the course of his phylogenesis;

Lastly, they are a zoological verticil, for, though in all appear-

[1] In present-day anthropoid apes, the cerebral capacity does not exceed 600 c.c.

ance narrowly localised on the farthest confines of Eastern Asia, this group obviously belonged to a very much bigger group whose nature and structure I shall be dealing with a little further on.

In short, *Pithecanthropus* and *Sinanthropus* are far from being merely a couple of interesting anthropological types. Through them, we are able to glimpse a whole wave of mankind.

Thus palaeontologists have once again shown their sense of proportion in picking out this very old and very primitive human layer and treating it as a distinct natural unity, to which they have even given a name, calling these early types the pre-hominids. This is an expressive and correct term from the standpoint of the anatomical progression of forms, but one liable to veil or misplace that psychic discontinuity in which we thought it necessary to place the very pith of hominisation. To call *Pithecanthropus* and *Sinanthropus* pre-hominids might suggest that they were not yet quite man. And that, according to my argument, would mean they had not yet crossed the threshold of reflection. The contrary seems to me much more probable; that, while admittedly far from having reached the level on which we stand, they were already, both of them, in the full sense of the word, intelligent beings.

That they were so seems to me to be stipulated by the general mechanism of phylogenesis. A mutation as fundamental as that of thought, a mutation which gives its specific impetus to the whole human group, could not in my opinion have appeared in the middle of the journey ; it could not have happened half-way up the stalk. It dominates the whole edifice. Its place must therefore be *beneath* every recognisable verticil in the unattainable depths of the peduncle, and thus beneath those creatures which (however pre-hominid in cranial structure) are already clearly situated *above* the point of origin and blossoming of our human race.

And there is more to it than that. So far we can find no trace of industry associated directly with the remains of *Pithecanthropus*. This is due to the conditions of where they lie :

around Trinil the fossils are of bones that have been carried down by streams to a lake. Near Peking, on the other hand, *Sinanthropus* has been caught in his lair, a filled-up cave littered with stone implements mixed with charred bones. Ought we, as M. Boule suggests, to see in this industry (sometimes, I admit, of an astonishing quality) the vestiges of *another* man, unknown to us, to whom *Sinanthropus*, himself not a *homo faber*, served as prey? As long as no remains are found of this hypothetical man, I consider the idea gratuitous and, everything considered, less scientific. *Sinanthropus* already worked stones and made fire. Until disproved, those two accomplishments must be considered on the same level as reflection, forming with it an integral part of the 'peduncle'. Taken together in one strand, these three elements crop up universally at the same time as mankind. That, objectively, is the situation.

And if it is really so, we see that despite their osteological features so reminiscent of the anthropoids, the pre-hominids were psychologically much nearer to us and thus phyletically much less young and primitive than might have been supposed. It must have taken time to discover fire and the art of making a cutting tool—so much so that there is plenty of room for at least one more human verticil still lower down, which we shall perhaps unearth one day in Villafranchian times.

We have already said that other hominids, at a similar stage of development, unquestionably lived at the same time as *Pithecanthropus* and *Sinanthropus*. Unfortunately we have only very inadequate relics of them: the famous jaw from Heidelberg perhaps, and the badly preserved cranium of *Africanthropus* from East Africa. This is not enough to enable us to work out the general physiognomy of the group. An observation may, however, serve to shed light indirectly on what we want to know.

We now know two species of *Pithecanthropus*, one relatively small, the other much more robust and 'brutal'. To these must be added two other forms positively gigantic, the one from Java represented by the fragment of a jaw, the other from South

China by some isolated teeth. This makes, with *Sinanthropus* (for the same period and the same continental fringe), five different types in all, certainly related.

This multiplicity of related forms living closely pressed together in a narrow strip, and also this curious common tendency to gigantism, surely suggest the idea of an isolated, marginal, zoological offshoot mutating upon itself in an almost autonomous manner. And so it might seem that what was going on in China and Malaya may have had its equivalent elsewhere, in the case of other stems farther west.

If this is so we should have to say that, zoologically speaking, the human group in the Lower Quaternary period still formed only a loosely coherent group in which the divergent structure, usual in animal verticils, was still dominant.

But at the same time, doubtless in the more central regions of the continents,[1] the elements of a new and more compact wave of mankind were mustering, ready to take over from this archaic world.

## 2. THE GROUP OF THE NEANDERTHALOIDS

Geologically, after the Lower Quaternary period, the curtain falls. During the interval, the Trinil deposits were folded ; the red earth of China was carved with valleys ready to receive their thick coating of yellow loess ; the face of Africa was further fissured ; elsewhere glaciations advanced and receded. When the curtain rises again some sixty thousand years ago, and we can see the scene again, we find that the pre-hominids have disappeared. Their place is now occupied by the Neanderthaloids.

This new humanity is much better represented by fossil remains than the preceding one, because it is not only more

[1] Perhaps among the populations whose anatomical form is still unknown, but whose 'bi-faced' industry can be followed in the ancient Pleistocene from the Cape to the Thames and from Spain to Java.

recent, but also more numerous. Little by little the network of thought has extended and consolidated.

We find both progress in number and progress in hominisation.

With *Pithecanthropus* and *Sinanthropus*, science could still hesitate, wondering what sort of creature it was dealing with. By the Middle Quaternary period, on the other hand, except for a moment's hesitation at the Spy cranium or the Neanderthal skull, there is never any serious doubt but that we are studying the vestiges of members of our own race. This great development of the brain, this industry of the caves, and for the first time those incontestable cases of burial—everything goes to show that we are in the presence of true man.

We have true man, then—but man who was not yet precisely us.

We find his cranium generally elongated, a low forehead, thick, prominent orbital ridges, a still noticeable prognathism of the face, as a rule the absence of canine fossae, absence of chin, large teeth without any distinct neck between crown and root. Confronted with these different features, no anthropologist could fail to recognise at a glance the fossil remains of a European Neanderthaloid. No people on earth today could be confused with him, not even an Australian Aborigine or an Aino. The advance from Trinil or Peking man is, as I have said, manifest ; but the gulf in relation to modern man is hardly less. Accordingly we have now another rung on the morphological ladder, another evolutionary stage. And in conformity with the laws of phylogenesis we must inevitably suspect another zoological verticil here, whose reality has not ceased to assert itself in pre-history in the last few years. When the first Mousterian crania were discovered in Western Europe, and when it became clear that these bones had not belonged either to idiots or degenerates, anatomists were naturally led to imagine that in the Middle Palaeolithic age the earth was peopled by men corresponding exactly to the Neanderthal type. Whence a certain disappointment, perhaps, when fresh discoveries, more and more numerous, failed to confirm this simple hypothesis. Actually the diversity of the Neanderthaloids, year by year more apparent, is precisely what we ought to have expected. For we can now see it is that very diversity which definitely gives to their 'bundle' its interest and its true physiognomy.

Of the forms called Neanderthaloids, our science today recognises two distinct groups at different levels of phyletic evolution, a group of terminal forms and an infant group :

*a.* *The terminal group.* Survivors, gradually dying out, of the more or less autonomous offshoots which probably composed the pre-hominid verticil—Solo man of Java, a direct and scarcely changed descendant of Trinil man,[1] in Africa the extraordinarily brutal Rhodesian man and, in Europe, unless I am mistaken, Neanderthal man himself who, in spite of his remarkable and persistent distribution over the whole of Western Europe, seems really to represent nothing but the last florescence of a dying stock.

*b.* *The infant group.* A nebulous, not easily distinguishable group of pseudo-Neanderthaloids with primitive features, but definitely modernised or modernisable—a rounder head, less prominent orbital ridges, canine fossae better marked, sometimes the beginning of a chin : such are Steinheim man and the finds in Palestine. They are incontestably Neanderthaloids, but they are ever so much nearer to us ; a progressive branch, sleeping, one might say, waiting for the coming dawn.

So let us put this triple 'bundle' in its proper light, geographically and morphologically. Far from being a disconcerting combination, the pattern is familiar. Leaves which have just fallen ; leaves still alive but beginning to turn yellow ; leaves not yet opened but full of vigour ; the complete section, almost an ideal one, of zoological ramification.

[1] Found in number in the horizontal terraces levelling the folded beds at Trinil, *homo soloensis* seems to have been simply a big *Pithecanthropus* with a more rounded cranium. This is an almost unique case in palaeontology, of one and the same phylum seen at the same place, across a geological discordance, at two different stages of its development.

## 3. THE *HOMO SAPIENS* COMPLEX

One of the great surprises of botany is to see at the beginning of the Cretaceous period the world of cycads and conifers abruptly submerged and replaced by a forest of angiosperms, plane trees, oaks, etc., the bulk of modern forms bursting ready-made on the Jurassic flora from some unknown region of the globe. No less is the anthropologist bewildered when he discovers, superimposed upon each other, hardly separated in the caves by a floor of stalagmites, Mousterian man and Cromagnon man or Aurignacian man. Here there is hardly any geological hiatus at all, yet none the less we find a fundamental rejuvenation of mankind. We find the sudden invasion of *Homo sapiens*, driven by climate or the restlessness of his soul, sweeping over the Neanderthaloids.

Where did he come from, this new man ? Some anthropologists would like to see in him the culmination of certain lines of development already pin-pointed in earlier epochs—a direct descendant, for example, of *Sinanthropus*. For definite technical reasons, however, and still more because of overall analogies, it is better to view things in another way. Without doubt, somewhere or other and *in his own way*, Upper Palaeolithic man must have passed through a pre-hominid phase and then through a Neanderthaloid one. But, like the mammals, the trituberculates, and all the other phyla, he disappears from our field of vision in the course of his (possibly accelerated) embryogenesis. We find imbrication and replacement rather than continuity and prolongation : the *law of succession* once again dominates history. I can thus easily picture the new-comer as the scion of an autonomous line of evolution, long hidden though secretly active—to emerge triumphantly one fine day doubtless in the midst of those pseudo-Neanderthaloids whose vital and probably very ancient ' bundle ' we have already mentioned. But at any rate, one thing is certain and admitted by everybody. The man we find on the face of the earth at the end of the Quaternary period is already modern man—and in every way.

First of all *anatomically* without any possible doubt. We see it in his high forehead with reduced orbits ; in his well-rounded parietal bones ; in his weak occipital crest now below his swelling brain ; in his slight jaw with its prominent chin—all these features, so well marked in the last cave-dwellers, are definitely our own. So clearly are they ours that, from this moment onwards, the palaeontologist, accustomed to working on pronounced morphological differences, no longer finds it easy to distinguish between the remains of these fossil men and men today. For that subtle task their over-all methods and visual sizing-up are no longer adequate, and they must now have recourse to the most delicate techniques (and audacities) of anthropology. We are no longer dealing with the reconstruction on general lines of the mounting horizons of life, but with the analysis of the overlapping nuances making up our foreground. Thirty thousand years. A long period measured in terms of our lifetime, but it is a mere second for evolution. From the osteological point of view there is during this interval no appreciable breach of continuity in the human phylum. It might even be said that there is, up to a point, no *major* change in the progress of its somatic ramification.

And this is where we get our greatest surprise. In itself, it is only very natural that the stem of *Homo sapiens fossilis*, studied at its point of emergence, far from being simple, should display in the composition and divergence of its fibres the complex structure of a fan. This is, as we know, the initial condition of each phylum on the tree of life. At the very least we should have counted on finding, in those depths, a cluster of relatively primitive and generalised forms, something antecedent in form to our present races. And what we find is rather the opposite. Assuming one can trust bones to give us an idea of flesh and blood, what were in fact those first representatives, in the age of the reindeer, of a new human verticil freshly opening ? Nothing more or less than what we see living today in approximately the same regions of the earth. Negroes, white men and yellow men (or at the most pre-negro, pre-white and pre-yellow), and those various groups already for the most part settled to north, to south, to east, to west, in their present geographical zones. That is what we find all over the ancient world from Europe to China at the end of the last Ice Age. Accordingly when we study Upper Palaeolithic man, not only in the essential features of his anatomy but also in the main lines of his ethnography, it is really ourselves and our own infancy that we are finding, not only the skeleton of modern man already there, but the framework of modern humanity. We see the same general bodily form ; the same fundamental distribution of races ; the same tendency (at least in outline) for the ethnic groups to join up together in a coherent system, over-riding all divergence. And (how could it fail to follow ?) the same essential aspirations in the depths of their soul.

Among the Neanderthaloids, as we have seen, a psychic advance was manifest, shown amongst other signs by the presence in the caves of the first graves. Even to the more brutal Neanderthals, everyone is prepared to grant the flame of a genuine intelligence. Most of it, however, seems to have been used up in the sheer effort to survive and reproduce. If there was any left over, we see no signs of it or fail to recognise them. What went on in the minds of those distant cousins of ours? We have no idea. But in the age of the reindeer, with *homo sapiens*, it is a definitely liberated thought which explodes, still warm, on to the walls of the caves. Within them, these new-comers brought art, an art still naturalistic but prodigiously accomplished. And thanks to the language of this art, we can for the first time enter right into the consciousness of these vanished beings whose bones we put together. There is a strange spiritual nearness, even in detail. Those rites expressed in red and black on the walls of caves in Spain, in the Pyrenees and Périgord, are after all still practised under our eyes in Africa, in Oceania, and even in America. What difference is there, for example, between the sorcerer of the Trois-Frères Cave dressed up in his deerskin, and some oceanic god ? But that's not the most important point. We could make mistakes in interpreting in modern terms the prints of hands, the bewitched bisons, and the fertility symbols which give expression to the preoccupation and religion of an Aurignacian or a Magdalenian man. Where we could not be mistaken is in perceiving in the artists of those distant ages a power of observation, a love of fantasy, and a joy in creation (manifest as much in the perfection of movement and outline as in the spontaneous play of chiselled ornament)—these flowers of a consciousness not merely reflecting upon itself, but rejoicing in so doing. So the examination of skeletons and skulls has not led us astray. In the Upper Quaternary period it is indeed and in the fullest sense present-day man at whom we are looking, not yet adult, admittedly, but having nevertheless reached the 'age of reason'. And when we compare him to ourselves, his brain is already perfect, so perfect that since that time there seems to have been no measurable variation or increased perfection in the organic instrument of our thought.

Are we to say, then, that the evolution in man ceased with the end of the Quaternary era ?

Not at all. But, without prejudice to what may still be developing slowly and secretly in the depths of the nervous system, evolution has since that date overtly overflowed its anatomical modalities to spread, or perhaps even to transplant its main thrust into the zones of psychic spontaneity both individual and collective.

Henceforward it is in that form almost exclusively that we shall be recognising it and following its course.

## 4. THE NEOLITHIC METAMORPHOSIS

Throughout living phyla, at all events among the higher animals where we can follow the process more easily, social development is a progress that comes relatively late. It is an achievement of maturity. In man, for reasons closely connected with his power of reflection, this transformation is accelerated. As far back as we can meet them, our great-great-ancestors are to be found *in groups* and gathered round the fire.

Definite as may be the signs of association at those remote periods, the whole phenomenon is far from being clearly outlined. Even in the Upper Palaeolithic era, the peoples we meet with seem to have constituted no more than loosely bound groups of wandering hunters. It was only in the Neolithic age that the great cementing of human elements began which was never thenceforward to stop. The Neolithic age, disdained by pre-historians because it is too young, neglected by historians because its phases cannot be exactly dated, was nevertheless a critical age and one of solemn importance among all the epochs of the past, for in it Civilisation was born.

Under what conditions did that birth take place ? Once again, and always in conformity with the laws regulating our vision of time in retrospect, we do not know. A few years ago it was usual to speak of a ' great gap ' between the last levels of chipped stone and the earliest levels of polished stone and pottery. Since then a series of intercalated horizons, better defined, have little by little brought together the verges of this gap, yet essentially the gulf still persists. Did it come from a play of migrations, or was it the effect of contagion ? Was it due to the sudden arrival of some ethnic wave, which had been silently assembling in some other and more fertile region of the globe, or the irresistible propagation of fruitful innovations ? Did the emphasis lie on a movement of peoples or primarily on a movement of cultures ? We should find it hard, as yet, to say. What is certain is that, after a gap geologically negligible, but long enough nevertheless for the selection and domestication of all the animals and plants on which we are still living today, we find sedentary and socially organised men in place of the nomadic hunters of the horse and the reindeer. In a matter of ten or twenty thousand years man divided up the earth and struck his roots in it.

In this decisive period of socialisation, as previously at the instant of reflection, a cluster of partially independent factors seems to have mysteriously converged to favour and even to force the pace of hominisation. Let us try to sort them out.

First of all come the incessant advances of multiplication. With the rapidly growing number of individuals the available land diminished. The groups pressed against one another. As a result migrations were on a smaller scale. The problem now was how to get the most out of ever more diminishing land, and we can well imagine that under pressure of this necessity the idea was born of conserving and reproducing on the spot what had hitherto been sought for and pursued far and wide. Agriculture and stock-breeding, the husbandman and the herdsman, replaced mere gathering and hunting.

From that fundamental change all the rest followed. In the growing agglomerations the complex of rights and duties began to appear, leading to the invention of all sorts of communal and juridical structures whose vestiges we can still see today in the shadow of the great civilisations among the least progressive populations of the world. In regard to property, morals and marriage, every possible social form seems to have been tried.

Simultaneously, in the more stable and more densely populated environment created by the first farms, the need and the taste for research were stimulated and became more methodical. It was a marvellous period of investigation and invention when, in the unequalled freshness of a new beginning, the eternal groping of life burst out in conscious reflection. Everything possible seems to have been attempted in this extraordinary period : the selection and empirical improvement of fruits, cereals, live-stock ; the science of pottery ; and weaving. Very soon followed the first elements of pictographic writing, and soon the first beginnings of metallurgy.

Then, in virtue of all this, consolidated on itself and better equipped for conquest, mankind could fling its final waves in the assault on those positions which had not yet fallen to it. Henceforward it was in the full flush of expansion. It was in fact at the dawn of the Neolithic age that man reached America (passing through an Alaska free of ice and perhaps by other ways) there to start again—on new material and at the cost of new efforts—his patient work of installation and domestication. Among them were many hunters and fishers still living a more or less Palaeolithic life despite their pottery and polished stone. But beside them were genuine tillers of the soil—the maize eaters. And at the same time, no doubt, another layer began to spread whose long trail is still marked by the presence of banana trees, mango trees and coconut palms—the fabulous adventure across the Pacific.

At the end of this metamorphosis (whose existence, once again, we can only just infer from the results) the world was practically covered with a population whose remains—polished stone implements, mill stones and shards, found under recent humus or sand deposits—litter the old earth of the continents.

Mankind was of course still very much split up. To get an idea of it, we must think of what the first white men found in America or Africa—a veritable mosaic of groups, profoundly different both ethnically and socially.

But mankind was already outlined and linked up. Since the age of the reindeer the peoples had been little by little finding their definitive place, even in matters of detail. Between them exchanges increased in the commerce of objects and the transmission of ideas. Traditions became organised and a collective memory was developed. Slender and granular as this first membrane might be, the noosphere there and then began to close in upon itself—and to encircle the earth.

## 5. THE PROLONGATIONS OF THE NEOLITHIC AGE AND THE RISE OF THE WEST

We have retained the habit, come down to us from the days when human palaeontology did not exist, of isolating that particular slice of six thousand years or so for which we possess written or dated documents. This for us is History, as opposed to pre-History. In reality, however, there is no breach of continuity between the two. The better we get the past into perspective, the more clearly we see that the periods called 'historic' (right down to and *including* the beginning of 'modern' times) are nothing else than direct prolongations of the Neolithic age. Of course, as we shall point out, there was increasing complexity and differentiation, but essentially following the same lines and on the *same plane*.

From the biological point of view—which is the one we are taking—how shall we define and represent the progress of hominisation in the course of this period, so short yet so prodigiously fruitful ?

Essentially, what history records among the welter of institutions, peoples and empires, is the normal expansion of *Homo sapiens* at the heart of the social atmosphere created by the Neolithic transformation. We find a gradual falling away of the oldest 'splinters' some of which, like the Australian aborigines, still adhere to the extreme fringe of our civilisation and our continents ; on the other hand we find accentuation and domination of certain other stems, more central and more vigorous, which attempt to monopolise the land and the light. Here and there we find disappearances causing a thinning-out, here and there some fresh buddings which make the foliage more dense. Some branches wither, some sleep, some shoot up and spread everywhere. We find endless interlacing of ramifications, none of which allow their peduncles to be seen clearly, not even at a mere two thousand years back ; in other words the whole series of cases, situations and appearances usually met with in any phylum in a state of active proliferation.

Nor is this quite all. We might suppose that, after the Neolithic age, what constituted the extreme difficulty, but also the exceptional interest, of human phylogenesis was the proximity of the facts, allowing us to follow with the naked eye, as it were, the biological mechanism of the ramification of the species. In fact, something more than that happens.

So long as science had to deal only with pre-historic human

groups, more or less isolated and to a greater or less extent undergoing anthropological formation, the general rules of animal phylogenesis were still approximately valid. From Neolithic times onwards the influence of psychical factors begins to outweigh—and by far—the variations of ever-dwindling somatic factors. And henceforward the foreground is taken up by the two series of effects we announced above when describing the main lines of hominisation—(i) the apparition above the genealogical verticils of political and cultural units ; a complex scale of groupings which, on the multiple planes of geographical distribution, economic links, religious beliefs and social institutions, have proved capable, after submerging 'the race', of reacting between themselves in every proportion ; and simultaneously (ii) the manifestation—between these branches of a new kind—of the forces of coalescence (anastomoses, confluences) liberated in each one by the individualisation of psychological sheath, or more precisely of an axis—a whole conjugated play of divergences and convergences.

There is no need for me to emphasise the reality, diversity and continual germination of human collective unities, at any rate potentially divergent ; such as the birth, multiplication and evolution of nations, states and civilisations. We see the spectacle on every hand, its vicissitudes fill the annals of the peoples. But there is one thing that must not be forgotten if we want to enter into and appreciate the drama. However hominised the events, the history of mankind in this rationalised form really does prolong—though in its own way and degree—the organic movements of life. It is *still* natural history through the phenomena of social ramification that it relates.

Much more subtle and fraught with biological potentialities are the phenomena of confluence. Let us try to follow them in their mechanism and their consequences.

Between animal branches or phyla of low 'psychical' endowment, reactions are limited to competition and eventually to elimination. The stronger supplants the weaker and ends by stifling it. The only exceptions to this brutal, almost mechanical law of substitution are those (mostly functional) associations of 'symbiosis' inferior organisms—or with the most socialised insects, the enslavement of one group by another.

With man (at all events with Post-Neolithic man) simple elimination tends to become exceptional, or at all events secondary. However brutal the conquest, the suppression is always accompanied by some degree of assimilation. Even when partially absorbed, the vanquished still reacts on the victor so as to transform him. There is, as the geologists call the process, endomorphosis—especially in the case of a peaceful cultural invasion, and yet still more with populations, equally resistant and active, which interpenetrate slowly under prolonged tension. What happens then is mutual permeation of the psychisms combined with a remarkable and significant interfecundity. Under this two-fold influence, veritable biological combinations are established and fixed which shuffle and blend ethnic traditions at the same time as cerebral genes. Formerly, on the tree of life we had a mere tangle of stems ; now over the whole domain of *Homo sapiens* we have synthesis.

But of course we do not find this everywhere to the same extent.

Because of the haphazard configuration of continents on the earth, some regions are more favourable than others for the concourse and mixing of races—extended archipelagoes, junctions of valleys, vast cultivable plains, particularly, irrigated by a great river. In such privileged places there has been a natural tendency ever since the installation of settled life for the human mass to concentrate, to fuse, and for its temperature to rise. Whence the no doubt 'congenital' appearance on the Neolithic layer of certain foci of attraction and organisation, the prelude and presage of some new and superior state for the noosphere. Five of these foci, of varying remoteness in the past, can easily be picked out—Central America, with its Maya civilisation ; the South Seas, with Polynesian civilisation ; the basin of Yellow River, with Chinese civilisation ; the valleys of the Ganges and the Indus, with Indian civilisation ; and lastly the

Nile Valley and Mesopotamia with Egyptian and Sumerian civilisation. The last three foci may have first appeared almost at the same period, the first two were much later. But they were all largely independent of one another, each struggling blindly to spread and ramify, as though it were alone destined to absorb and transform the earth.

Basically can we not say that the essential thing in history consists in the conflict and finally the gradual harmonisation of these great psycho-somatic currents ?

In fact this struggle for influence was quickly localised. The Maya centre which was too isolated in the New World, and the Polynesian centre which was too dispersed on the monotonous dust of its distant islands, soon met their respective fates, one being completely extinguished and the other radiating in a vacuum. So finally the contest for the future of the world was fought out by the agricultural plain dwellers of Asia and North Africa. One or two thousand years before our era the odds between them may have seemed fairly equal. But we today, in the light of events, can see that even at that stage there were the seeds of weakness in two of the contestants in the East.

Either by its own genius or as an effect of immensity, China (and I mean the *old* China, of course) lacked both the inclination and the impetus for deep renovation. A singular spectacle is presented by this gigantic country which only yesterday represented—still living under our eyes—a scarcely changed fragment of the world as it could have been ten thousand years ago. The population was not only fundamentally agricultural but essentially organised according to the hierarchy of territorial possessions—the emperor being nothing more than the biggest proprietor. It was a population ultra-specialised in brick work, pottery and bronze, a population carrying to the lengths of superstition the study of pictograms and the science of the constellations ; an incredibly refined civilisation, admittedly, but unchanged as to method since its beginning, like the writing which betrays the fact so ingenuously. Well into the nineteenth century it was still Neolithic, not rejuvenated, as elsewhere, but simply interminably complicated in on itself, not merely continuing on the same lines, but remaining on the same level, as though unable to life itself above the soil where it was formed.

And while China, already encrusted in its soil, multiplied its gropings and discoveries without ever taking the trouble to build up a science of physics, India allowed itself to be drawn into metaphysics, only to become lost there. India—the region *par excellence* of high philosophic and religious pressures : we can never make too much of our indebtedness to the mystic influences which have come down to each and all of us in the past from this 'anticyclone'. But however efficacious these currents for ventilating and illuminating the atmosphere of mankind, we have to recognise that, with their excessive passivity and detachment, they were incapable of building the world. The primitive soul of India arose in its hour like a great wind but, like a great wind also, again in its hour, it passed away. How indeed could it have been otherwise ? Phenomena regarded as an illusion (Maya) and their connections as a chain (Karma), what was left in these doctrines to animate and direct human evolution ? A simple mistake was made—but it was enough—in the definition of the spirit and in the appreciation of the bonds which attach it to the sublimations of matter.

Then step by step we are driven nearer to the more western zones of the world—to the Euphrates, the Nile, the Mediterranean—where an exceptional concurrence of places and peoples was, in the course of a few thousand years, to produce that happy blend, thanks to which reason could be harnessed to facts and religion to action. And this without losing any of their upward thrust—in fact quite the contrary. Mesopotamia, Egypt, Greece—with Rome soon to be added—and above all the mysterious Judaeo-Christian ferment which gave Europe its spiritual form. But I shall be coming back to that at the end of this book.

It is easy for the pessimist to reduce this extraordinary period to a number of civilisations which have fallen into ruins one after the other. Is it not far more scientific to recognise, yet once again, beneath these successive oscillations, the great spiral of life : thrusting up, irreversibly, in relays, following the master-line of its evolution ? Susa, Memphis and Athens can crumble. An ever more highly organised consciousness of the universe is passed from hand to hand, and glows steadily brighter.

Later on, when I come to speak of the current planetisation of the noosphere, I shall try to restore to the other fragments of mankind the great and essential part reserved for them in the expected plenitude of the earth. At this point of our investigation, we would be allowing sentiment to falsify the facts if we failed to recognise that during historic time the principal axis of anthropogenesis has passed through the West. It is in this ardent zone of growth and universal recasting that all that goes to make man today has been discovered, or at any rate *must have been rediscovered*. For even that which had long been known elsewhere only took on its definitive human value in becoming incorporated in the system of European ideas and activities. It is not in any way naïve to hail as a great event the discovery by Columbus of America.

In truth, a neo-humanity has been germinating round the Mediterranean during the last six thousand years, and precisely at this moment it has finished absorbing the last vestiges of the Neolithic mosaic; thus starts the budding of another layer on the noosphere, and the densest of all.

The proof of this lies in the fact that from one end of the world to the other, all the peoples, to remain human or to become more so, are inexorably led to formulate the hopes and problems of the modern earth in the very same terms in which the West has formulated them.

CHAPTER THREE

# THE MODERN EARTH

## *A Change of Age*

IN EVERY epoch man has thought himself at a 'turning-point of history'. And to a certain extent, as he is advancing on a rising spiral, he has not been wrong. But there are moments when this impression of transformation becomes accentuated and is thus particularly justified. And we are certainly not exaggerating the importance of our contemporary existences in estimating that, upon them, a turn of profound importance is taking place in the world which may even crush them.

When did this turn begin ? It is naturally impossible to say exactly. Like a great ship, the human mass only changes its course gradually, so much so that we can put far back—at least as far as the Renaissance—the first vibrations which indicate the change of route. It is clear, at any rate, that at the end of the eighteenth century the course had been changed in the West. Since then, in spite of our occasional obstinacy in pretending that we are the same, we have in fact entered a different world.

Firstly, economic changes. Advanced as it was in many ways two centuries ago, our civilisation was still based fundamentally on the soil and its partition. The type of 'real' property, the nucleus of the family, the prototype of the state (and even the universe) was still, as in the earliest days of society, the arable field, the territorial basis. Then, little by little, as a result of the 'dynamisation' of money, property has evaporated into something fluid and impersonal, so mobile that already the wealth of nations themselves has almost nothing in common with their frontiers.

Secondly, industrial changes. Up to the eighteenth century, in spite of the many improvements made, there was still only one known source of chemical energy—fire. And there was only one sort of mechanical energy employed—muscle, human or animal, multiplied by the machine.

Lastly, social changes and the awakening of the masses.

Merely from looking at these external signs we can hardly fail to suspect that the great unrest which has pervaded our life in the West ever since the storm of the French Revolution springs from a nobler and deeper cause than the difficulties of a world seeking to recover some ancient equilibrium that it has lost. There is no question of shipwreck. What we are up against is the heavy swell of an unknown sea which we are just entering from behind the cape that protected us. What is troubling us intellectually, politically and even spiritually is something quite simple. With his customary acute intuition, Henri Breuil said to me one day: 'We have only just cast off the last moorings which held us to the Neolithic age.' The formula is paradoxical but illuminating. In fact the more I have thought over these words, the more inclined I have been to think that Breuil was right.

We are, at this very moment, passing through a change of age.

The age of industry; the age of oil, electricity and the atom; the age of the machine, of huge collectivities and of science—the future will decide what is the best name to describe the era we are entering. The word matters little. What does matter is that we should be told that, at the cost of what we are enduring, life is taking a step, and a decisive step, in us and in our environment. After the long maturation that has been steadily going on during the apparent immobility of the agricultural centuries, the hour has come at last, characterised by the birth pangs inevitable in another change of state. There were the first men—those who witnessed our origin. There are others who will witness the great scenes of the end. To us, in our brief span of life, falls the honour and good fortune of coinciding with a critical change of the noosphere.

In these confused and restless zones in which present blends with future in a world of upheaval, we stand face to face with all the grandeur, the unprecendented grandeur, of the phenomenon of man. Here if anywhere, now if ever, have we, more legitimately than any of our predecessors, the right to think that we can measure the importance and detect the direction of the process of hominisation. Let us look carefully and try to understand. And to do so let us probe beneath the surface and try to decipher the particular form of mind which is coming to birth in the womb of the earth today.

Our earth of factory chimneys and offices, seething with work and business, our earth with a hundred new radiations—this great organism lives, in final analysis, only because of, and for the sake of, a new soul. Beneath a change of age lies a change of thought. Where are we to look for it, where are we to situate this renovating and subtle alteration which, without appreciably changing our bodies, has made new creatures of us ? In one place and one only—in a new intuition involving a total change in the physiognomy of the universe in which we move—in other words, in an awakening.

What has made us in four or five generations so different from our forebears (in spite of all that may be said), so ambitious too, and so worried, is not merely that we have discovered and mastered other forces of nature. In final analysis it is, if I am not mistaken, that we have become conscious of the movement which is carrying us along, and have thereby realised the formidable problems set us by this reflective exercise of the human effort.

## 1. THE DISCOVERY OF EVOLUTION

### A. *The Perception of Space-time*[1]

We have all forgotten the moment when, opening our eyes for the first time, we saw light and things around us all jumbled up and all on one single plane. It requires a great effort to imagine the time when we were unable to read or again to take our minds back to the time when for us the world extended no farther than the walls of our home and our family circle.

Similarly it seems to us incredible that men could have lived without suspecting that the stars are hung above us hundreds of light years away, or that the contours of life stretched out millions of years behind us to the limits of our horizon. Yet we have only to open any of those books with barely yellowing pages in which the authors of the sixteenth, or even as late as the eighteenth, century discoursed on the structure of worlds to be startled by the fact that our great-great-great-grandfathers felt perfectly at ease in a cubic space where the stars turned round the earth, and had been doing so for less than 6,000 years. In a cosmic atmosphere which would suffocate us from the first moment, and in perspectives in which it is physically impossible for us to enter, they breathed without any inconvenience, if not very deeply.

Between them and us what, then, has happened ?

I know of no more moving story nor any more revealing of the biological reality of a noogenesis than that of intelligence struggling step by step from the beginning to overcome the encircling illusion of proximity.

In the course of this struggle to master the dimensions and the relief of the universe, space was the first to yield—naturally, because it was more tangible. In fact the first hurdle was taken in this field when long, long ago a man (some Greek, no doubt, before Aristotle), bending back on itself the apparent flatness of

[[1] Cf. Collingwood, *Idea of Nature* (O.U.P. 1944) *passim.*]

things, had an intuition that there were antipodes. From then onwards round the round earth the firmament itself rolled roundly. But the focus of the spheres was badly placed. By its situation it incurably paralysed the elasticity of the system. It was only really in the time of Galileo, through rupture with the ancient geocentric view, that the skies were made free for the boundless expansions which we have since detected in them. The earth became a mere speck of sidereal dust. Immensity became possible, and to balance it the infinitesimal sprang into existence.

For lack of apparent yardsticks, the depths of the past took much longer to be plumbed. The movement of stars, the shape of mountains, the chemical nature of bodies—indeed all matter seemed to express a continual present. The physics of the seventeenth century was incapable of opening Pascal's eyes to the abysses of the past. To discover the real age of the earth and then of the elements, it was necessary for man to become fortuitously interested in an object of moderate mobility, such as life, for instance, or even volcanoes. It was thus through a narrow crack (that of 'natural history', then in its infancy) that from the eighteenth century onwards light began to seep down into the great depths beneath our feet. In these initial estimates, the time considered necessary for the formation of the world was still very modest. But at least the impetus had been given and the way out opened up. After the walls of space, shaken by the Renaissance, it was the floor (and consequently the ceiling) of time which, from Buffon onwards, became mobile. Since then, under the unceasing pressure of facts, the process has continually accelerated. Although the strain has been taken off for close on two hundred years, the spirals of the world have still not been relaxed. The distance between the turns in the spiral has seemed ever greater and there have always been further turns appearing deeper still.

Yet in these first stages in man's awakening to the immensities of the cosmos, space and time, however vast, still remained homogeneous and independent of each other; they were two great containers, quite separate one from the other, extending infinitely no doubt, but in which things floated about or were packed together in ways owing nothing to the nature of their setting.

The two compartments had been enlarged beyond measure, but within each of them the objects seemed as freely transposable as before. It seemed as if they could be placed here or there, moved forward, pushed back or even suppressed at will. If no-one ventured formally as far as this play of thought, at least there was still no clear idea why or to what extent it was impossible. This was a question which did not arise.

It was only in the middle of the nineteenth century, again under the influence of biology, that the light dawned at last, revealing the *irreversible coherence* of all that exists. First the concatenations of life and, soon after, those of matter. The least molecule is, in nature and in position, a function of the whole sidereal process, and the least of the protozoa is structurally so knit into the web of life that its existence cannot be hypothetically annihilated without *ipso facto* undoing the whole network of the biosphere. The *distribution, succession and solidarity of objects are born from their concrescence in a common genesis.* Time and space are organically joined again so as to weave, together, the stuff of the universe. That is the point we have reached and how we perceive things today.

Psychologically what is hidden behind this initiation ? One might well become impatient or lose heart at the sight of so many minds (and not mediocre ones either) remaining today still closed to idea of evolution, if the whole of history were not there to pledge to us that a truth once seen, even by a single mind, always ends up by imposing itself on the totality of human consciousness. For many, evolution is still only transformism, and transformism is only an old Darwinian hypothesis as local and as dated as Laplace's conception of the solar system or Wegener's Theory of Continental Drift. Blind indeed are those who do not see the sweep of a movement whose orbit infinitely transcends the natural sciences and has successively invaded and conquered the surrounding territory—chemistry, physics, sociology and even mathematics and the history of religions. One after the other all the fields of human knowledge have been shaken and carried away by the same under-water current in the direction of the study of some *development*. Is evolution a theory, a system or a hypothesis? It is much more: it is a general condition to which all theories, all hypotheses, all systems must bow and which they must satisfy henceforward if they are to be thinkable and true. Evolution is a light illuminating all facts, a curve that all lines must follow.

In the last century and a half the most prodigious event, perhaps, ever recorded by history since the threshold of reflection has been taking place in our minds: the definitive access of consciousness to a *scale of new dimensions*; and in consequence the birth of an entirely renewed universe, without any change of line or feature by the simple transformation of its intimate substance.

Until that time the world seemed to rest, static and fragmentable, on the three axes of its geometry. Now it is a casting from a single mould.

What makes and classifies a 'modern' man (and a whole host of our contemporaries is not yet 'modern' in this sense) is having become capable of seeing in terms not of space and time alone, but also of duration, or—it comes to the same thing—of biological space-time; and above all having become incapable of seeing anything otherwise—anything—*not even himself*.

This last step brings us to the heart of the metamorphosis.

## B. *The Envelopment in Duration*

Obviously man could not see evolution all around him without feeling to some extent carried along by it himself. Darwin has demonstrated this. Nevertheless, looking at the progress of transformist views in the last hundred years, we are surprised to see how naïvely naturalists and physicists were able at the early stages to imagine themselves to be standing outside the universal stream they had just discovered. Almost incurably subject and object tend to become separated from each other in the act of knowing. We are continually inclined to isolate ourselves from the things and events which surround us, as though we were looking at them from outside, from the shelter of an observatory into which they were unable to enter, as though we were spectators, not elements, in what goes on. That is why, when it was raised by the concatenations of life, the question of man's origins was for so long restricted to the purely somatic and bodily side. A long animal heredity might well have formed our limbs, but our mind was always above the play of which it kept the score. However materialistic they might be, it did not occur to the first evolutionists that their scientific intelligence had anything to do in itself with evolution.

At this stage they were only half-way to the truth they had discovered.

From the very first pages of this book, I have been relentlessly insisting on one thing : for invincible reasons of homogeneity and coherence, the fibres of cosmogenesis demand their prolongation in us in a way that goes far deeper than flesh and blood. We are not only set adrift and carried away in the current of life by the material surface of our being ; but, like a subtle fluid, space-time first drowns our bodies and then penetrates to our soul ; it fills it and impregnates it ; it blends itself with the soul's potentialities to such an extent that soon the soul no longer knows how to distinguish space-time from its own thoughts. To those who can use their eyes nothing, not even at the summit of our being, can escape this flux any longer, because it is only definable in increase of consciousness. The very act by which the fine edge of our minds penetrates the absolute is a phenomenon, as it were, of *emergence*. In short, first recognised only at a single point, then perforce extended to the whole inorganic and organic volume of matter, evolution is now, whether we like it or not, gaining the psychic zones of the world and transferring to the spiritual constructions of life not only the cosmic stuff but also the cosmic 'primacy' hitherto reserved by science to the tangled whirlwind of the ancient 'ether'.

How indeed could we incorporate thought into the organic flux of space-time without being forced to grant it the first place in the processus? How could we imagine a cosmogenesis reaching right up to mind without being thereby confronted with a noogenesis?

Thus we see not only thought as participating in evolution as an anomaly or as an epiphenomenon; but evolution as so reducible to and identifiable with a progress towards thought that the movement of our souls expresses and measures the very stages of progress of evolution itself. Man discovers that *he is nothing else than evolution become conscious of itself*, to borrow Julian Huxley's striking expression. It seems to me that our modern minds (because and inasmuch as they are modern) will never find rest until they settle down to this view. On this summit and on this summit alone are repose and illumination waiting for us.

### C. *The Illumination*

The consciousness of each of us is evolution looking at itself and reflecting upon itself.

With that very simple view, destined, as I suppose, to become as instinctive and familiar to our descendants as the discovery of a third dimension in space is to a baby, a new light—inexhaustibly harmonious—bursts upon the world, radiating from ourselves.

Step by step, from the early earth onwards, we have followed *going upwards* the successive advances of consciousness in matter undergoing organisation. Having reached the peak, we can now turn round and, *looking downwards*, take in the pattern of the whole. And this second check is decisive, the harmony is perfect. From any other point of view, there is always a 'snag': something clashes, for there is no natural place no genetic place—for human thought in the landscape. Whereas here, from top to bottom, from our souls and *including* our souls, the lines stretch in both directions, untwisted and unbroken. From top to bottom, a triple unity persists and develops: unity of structure, unity of mechanism and unity of movement.

*a. Unity of structure.* 'Verticils' and 'fannings out'.

On every scale, this is the pattern we see on the tree of life. We found it again at the origins of mankind and of the principal human waves. We have seen it with our own eyes today in the complex ramifications of nations and races. And now, with an eye rendered more sensitive by training, we shall be able to discern the same pattern again in forms which are more and more immaterial and near.

Our habit is to divide up our human world into compartments of different sorts of 'realities': natural and artificial, physical and moral, organic and juridical, for instance.

In a space-time, legitimately and perforce extended to include the movements of the mind within us, the frontiers between these pairs of opposites tend to vanish. Is there after all such a great difference from the point of view of the expansion of life between a vertebrate either spreading its limbs or equipping them with feathers, and an aviator soaring on wings with which he has had the ingenuity to provide himself? In what way is the ineluctable play of the energies of the heart less physically real than the principle of universal attraction? And, conventional and impermanent as they may seem on the surface, what are the intricacies of our social forms, if not an effort to isolate little by little what are one day to become the structural laws of the noosphere? In their essence, and provided they keep their vital connection with the current that wells up from the depths of the past, are not the artificial, the moral and the juridical simply the hominised versions of the natural, the physical and the organic?

From this point of view, which is that of the future natural history of the world, distinctions we cling to from habit (at the risk of over-partitioning the world) lose their value. Hence the ramifications of evolution reappear and go on close to us in a thousand social phenomena which we should never have imagined to be so closely linked with biology ; in the formation and dissemination of languages, in the development and specialisation of new industries, in the formulation and propagation of philosophic and religious doctrines. In each of these groups of human activity a superficial glance would only detect a weak and haphazard reproduction of the procedures of life. It would accept without questioning the strange fact of parallelism—or it would account verbally for it in terms of some abstract necessity.

For a mind that has awakened to the full meaning of evolution, mere inexplicable similitude is resolved in identity—the identity of a structure which, under different forms, extends from the bottom to the top, from threshold to threshold, from the roots to the flowers—by the organic continuity of movement or, which amounts to the same thing, by the organic unity of *milieu*.

*The social phenomenon is the culmination and not the attenuation of the biological phenomenon.*

*b. Unity of mechanism.* ' Groping ' and ' invention '.

It was to these words that we turned instinctively when we ran up against the facts of ' mutations ' in describing the appearance of successive zoological groups.

What exactly are these words worth, imbued as they may well be with anthropomorphism ?

Mutation reappears undeniably at the origin of the ramifications of institutions and ideas which interlace to form human society. Everywhere around us it is constantly cropping up, and precisely under the two forms that biology has divined and between which it hesitates : on the one hand we have mutations narrowly limited round a single focus ; on the other ' mass mutations ' in which whole blocks of mankind are swept along as by a flood. Here, however, because the phenomenon takes place in ourselves with its procedure in full view, we cannot be mistaken : we can see that in interpreting the progressive leaps of life in an active and finalist way we are not in error. For if our ' artificial ' constructions are really nothing but the legitimate sequel to our phylogenesis, *invention* also—this revolutionary act from which the creations of our thought emerge one after the other—can legitimately be regarded as an extension in reflective form of the obscure mechanism whereby each new form has always germinated on the trunk of life.

This is no metaphor, but an analogy founded in nature. We find the same thing in both—only it is easier to define in the hominised state.

And so, here again, we find that light reflected on itself, glancing off and in a flash descending to the lowest frontiers of the past. But this time what its beam illuminates in us at our lowest stages is no longer an endless play of tangled verticils, but a long sequence of discoveries. In the same beam of light the instinctive gropings of the first cell link up with the learned gropings of our laboratories. So let us bow our heads with respect for the anxieties and joys of ' trying all and discovering all '. The passing wave that we can feel was not formed in ourselves. It comes to us from far away ; it set out at the same time as the light from the first stars. It reaches us after creating everything on the way. The spirit of research and conquest is the permanent soul of evolution.

*c.* And hence, throughout all time, *unity of movement.* ' The rise and expansion of consciousness.'

Man is not the centre of the universe as once we thought in our simplicity, but something much more wonderful—the arrow pointing the way to the final unification of the world in terms of life. Man alone constitutes the last-born, the freshest, the most complicated, the most subtle of all the successive layers of life.

This is nothing else than the fundamental vision and I shall leave it at that.

But this vision, mind you, only acquires its full value—is indeed only defensible—through the simultaneous illumination within ourselves of the laws and conditions of heredity.

As I have already had occasion to say, we do not yet know how characters are formed, accumulated and transmitted in the secret recesses of the germ cells. Or rather, so long as it is talking of plants and animals, biology has not yet found a way of reconciling in phylogenesis the spontaneous activity of individuals with the blind determinism of the genes. In its inability to do so it is inclined to make the living being the passive and powerless witness of the transformations he undergoes—without being able to influence them and without being responsible for them.

But then (and this is the moment to settle the question once and for all), in the phylogenesis of mankind, what becomes of the part, obvious enough, played by the power of invention ?

What evolution perceives of itself in man by reflecting itself in him is enough to dispel or at least to correct these paradoxical appearances.

Certainly in our innermost being we all feel the weight, the stock of obscure powers, good or bad, a sort of definite and unalterable ' quantum ' handed down to us once and for all from the past. But with no less clarity we see that the further advance of the vital wave beyond us depends on how industriously we use those powers. How could we doubt this when we see them directly before us, through all the channels of ' tradition ', stored up irreversibly in the highest form of life accessible to our experience—I mean the collective memory and intelligence of the human biota? Ever under the influence of our tendency to disparage the ' artificial ', we are apt to regard these social functions—tradition, education and upbringing—as pale images, almost parodies, of what takes place in the natural formation of species. If the noosphere is not an illusion, is it not much more exact to recognise in these communications and exchanges of ideas the higher form, in which they come to be fixed in us, of the less supple modes of biological enrichments by *additivity*?

In short, the further the living being emerges from the anonymous masses by the radiation of his own consciousness, the greater becomes the part of his activity which can be stored up and transmitted by means of education and imitation. From this point of view man only represents an extreme case of transformation. Transplanted by man into the thinking layer of the earth, heredity, without ceasing to be germinal (or chromosomatic) in the individual, finds itself, by its very life-centre, settled in a reflecting organism, collective and permanent, in which phylogenesis merges with ontogenesis. From the chain of cells it passes into the circumterrestrial layers of the noosphere. There is nothing surprising if from this moment onwards, and thanks to the characters of this new *milieu*, it is reduced in its finest part to the pure and simple transmission of *acquired* spiritual treasures.

Passive as it may have been before reflection, heredity now springs to life, supremely active, in its noospheric form—that is to say, by becoming hominised.

Hence we were not saying enough when we said that evolution, by becoming conscious of itself in the depths of ourselves, only needs to look at itself in the mirror to perceive itself in all its depths and to decipher itself. In addition it becomes free to dispose of itself—it can give itself or refuse itself. Not only do we read in our slightest acts the secret of its proceedings ; but for an elementary part *we hold it in our hands*, responsible for its past to its future.

Is this grandeur or servitude ? Therein lies the whole problem of action.

## 2. THE PROBLEM OF ACTION

### A. *Modern Disquiet*

It is impossible to accede to a fundamentally new environment without experiencing the inner terrors of a metamorphosis. The child is terrified when it opens its eyes for the first time. Similarly, for our mind to adjust itself to lines and horizons enlarged beyond measure, it must renounce the comfort of familiar narrowness. It must create a new equilibrium for everything that had formerly been so neatly arranged in its small inner world. It is dazzled when it emerges from its dark prison, awed to find itself suddenly at the top of a tower, and it suffers from giddiness and disorientation. The whole psychology of modern disquiet is linked with the sudden confrontation with space-time.

It cannot be denied that, in a primordial form, human anxiety is bound up with the very advent of reflection and is thus as old as man himself. Nor do I think that anyone can seriously doubt the fact that, under the influence of reflection undergoing socialisation, the men of today are particularly uneasy, more so than at any other moment of history. Conscious or not, anguish —a fundamental anguish of being—despite our smiles, strikes in the depths of all our hearts and is the undertone of all our conversations. This does not mean that its cause is clearly recognised —far from it. Something threatens us, something is more than ever lacking, but without our being able to say exactly what.

Let us try then, step by step, to localise the source of our disquiet, eliminating the illegitimate causes of disturbance till we find the exact site of the pain at which the remedy, if there is one, should be applied.

In the first and most widespread degree, the 'malady of space-time' manifests itself as a rule by a feeling of futility, of being crushed by the enormities of the cosmos.

The enormity of space is the most tangible and thus the most frightening aspect. Which of us has ever in his life really had the courage to look squarely at and try to 'live' a universe formed of galaxies whose distance apart runs into hundreds of thousands of light years ? Which of us, having tried, has not emerged from the ordeal shaken in one or other of his beliefs ? And who, even when trying to shut his eyes as best he can to what the astronomers implacably put before us, has not had a confused sensation of a gigantic shadow passing over the serenity of his joy ?

Enormity of duration—sometimes having the effect of an abyss on those few who are able to see it, and at other times more usually (on those whose sight is poor), the despairing effect of stability and monotony. Events that follow one another in a circle, vague pathways which intertwine, leading nowhere.

Corresponding enormity of number—the bewildering number

of all that has been, is, and will be necessary to fill time and space. An ocean in which we seem to dissolve all the more irresistibly the more lucidly alive we are. The effort of trying conscientiously to find our proper place among a thousand million men. Or merely in a crowd.

Malady of multitude and immensity . . .

To overcome this first form of its uneasiness, I believe that the modern world has no choice but to proceed unhesitatingly right to the end of its intuition.

As motionless or blind (and by that I mean so long as we think of them as motionless or blind) time and space are indeed terrifying. Accordingly what could make our initiation into the true dimensions of the world dangerous is for it to remain incomplete, deprived of its complement and necessary corrective —the perception of an evolution animating those dimensions. On the other hand, what matters the giddy plurality of the stars and their fantastic spread, if that immensity (symmetrical with the infinitesimal) has no other function but to equilibrate the intermediary layer where, and where only in the medium range of size, life can build itself up chemically ? What matter the millions of years and milliards of beings that have gone before if those countless drops form a current that carries us along ? Our consciousness would evaporate, as though annihilated, in the limitless expansions of a static or endlessly moving universe. It is inwardly reinforced in a flux which, incredibly vast as it may be, is not only *becoming* but *genesis*, which is something quite different. Indeed time and space become humanised as soon as a definite movement appears which gives them a physiognomy.

'There is nothing new under the sun' say the despairing. But what about you, O thinking man ? Unless you repudiate reflection, you must admit that you have climbed a step higher than the animals. 'Very well, but at least nothing has changed and nothing is changing any longer since the beginning of history.' In that case, O man of the twentieth century, how does it happen that you are waking up to horizons and are susceptible to fears that your forefathers never knew ?

In truth, half our present uneasiness would be turned into happiness if we could once make up our minds to accept the facts and place the essence and the measure of our modern cosmogonies within a noogenesis. Along this axis no doubt is possible. The universe has always been in motion and at this moment continues to be in motion. But will it still be in motion *tomorrow* ?

Here only, at this turning point where the future substitutes itself for the present and the observations of science should give way to the anticipations of a faith, do our perplexities legitimately and indeed inevitably begin. Tomorrow ? But who can guarantee us a tomorrow anyway ? And without the assurance that this tomorrow exists, can we really go on living, we to whom has been given—perhaps for the first time in the whole story of the universe—the terrible gift of foresight ?

Sickness of the dead end—the anguish of feeling shut in . . .

This time we have at last put our finger on the tender spot.

What makes the world in which we live specifically modern is our discovery in it and around it of evolution. And I can now add that what disconcerts the modern world at its very roots is not being sure, and not seeing how it ever could be sure, that there is an outcome—*a suitable outcome*—to that evolution.

Now what should the future be like in order to give us the strength or even the joy to accept the prospect of it and bear its weight ?

To come to grips with the problem and see if there is a remedy, let us examine the whole situation.

### B. *The Requirements of the Future*

There was a time when life held sway over none but slaves and children. To advance, all it needed was to feed obscure instincts —the bait of food, the urge of reproduction, the half-confused struggle for a place in the sun, stepping over others, trampling them down if need be. The aggregate rose automatically and docile, as the resultant of an enormous sum of egoisms given rein. There was a time too, almost within living memory, when the workers and the disinherited accepted without reflection the lot which kept them in servitude to the remainder of society.

Yet when the first spark of thought appeared upon the earth, life found it had brought into the world a power capable of criticising it and judging it. This formidable risk which long lay dormant, but whose dangers burst out with our first awakening to the idea of evolution. Like sons who have grown up, like workers who have become ' conscious ', we are discovering that something is developing in the world by means of us, perhaps at our expense. And what is more serious still is that we have become aware that, in the great game that is being played, we are the players as well as being the cards and the stakes. Nothing can go on if we leave the table. Neither can any power force us to remain. Is the game worth the candle, or are we simply its dupes? This question has hardly been formulated as yet in man's heart, accustomed for hundreds of centuries to toe the line ; it is a question, however, whose mere murmur, already audible, infallibly predicts future rumblings. The last century witnessed the first systematic strikes in industry ; the next will surely not pass without the threat of strikes in the noosphere.

There is a danger that the elements of the world should refuse to serve the world—because they think ; or more precisely that the world should refuse itself when perceiving itself through reflection. Under our modern disquiet, what is forming and growing is nothing less than an organic crisis in evolution.

And now, at what price and on what contractual bases will order be restored ? On all the evidence, that is the nub of the problem.

In the critical disposition of mind we shall be in from now on, one thing is clear. We shall never bend our backs to the task that has been allotted us of pushing noogenesis onward except on condition that the effort demanded of us has a chance of succeeding and of taking us as far as possible. An animal may rush headlong down a blind alley or towards a precipice. Man will never take a step in a direction he knows to be blocked. There lies precisely the ill that causes our disquiet.

Having got so far, what are the minimum requirements to be fulfilled before we can say that the road ahead of us is *open* ? There is only one, but it is everything. It is that we should be assured the space and the chances to fulfil ourselves, that is to say, to progress till we arrive (directly or indirectly, individually or collectively) at the utmost limits of ourselves. This is an elementary request, a basic wage, so to speak, veiling nevertheless a stupendous demand. But is not the end and aim of thought that still unimaginable farthest limit of a convergent sequence, propagating itself without end and ever higher ? Does not the end or confine of thought consist precisely in not having a confine ? Unique in this respect among all the energies of the universe, consciousness is a dimension to which it is inconceivable and even contradictory to ascribe a ceiling or to suppose that it can double back upon itself. There are innumerable critical points on the way, but a halt or a reversion is impossible, and for the simple reason that every increase of internal vision is essentially the germ of a further vision which includes all the others and carries still farther on.

Hence this remarkable situation—that our mind, by the very fact of being able to discern infinite horizons ahead, is only able to move by the hope of achieving, through something of itself, a supreme consummation—without which it would rightly feel itself to be stunted, frustrated and cheated. By the nature of the work, and correlatively by the requirement [*exigence*] of the worker, a total death, an unscalable wall, on which consciousness would crash and then for ever disappear, are thus 'incompossible' with the mechanism of conscious activity (since it would immediately break its mainspring).

The more man becomes man, the less will he be prepared to move except towards that which is interminably and indestructibly new. Some ' absolute ' is implied in the very play of his operative activity.

After that, ' positive and critical ' minds can go on saying as much as they like that the new generation, less ingenuous than their elders, no longer believes in a future and in a perfecting of the world. Has it even occurred to those who write and repeat these things that, if they were right, all spiritual movement on earth would be virtually brought to a stop ? They seem to believe that life would continue its peaceful cycle when deprived of light, of hope and of the attraction of an inexhaustible future. And this is a great mistake. Flowers and fruit might still go on perhaps for a few years more by habit. But from these roots the trunk would be well and truly severed. Even on stacks of material energy, even under the spur of immediate fear or desire, *without the taste for life*, mankind would soon stop inventing and constructing for a work it knew to be doomed in advance. And, stricken at the very source of the impetus which sustains it, it would disintegrate from nausea or revolt and crumble into dust.

Having once known the taste of a universal and durable progress, we can never banish it from our minds any more than our intelligence can escape from the space-time perspective it once has glimpsed.

If progress is a myth, that is to say, if faced by the work involved we can say : ' What's the good of it all ? ' our efforts will flag. With that the whole of evolution will come to a halt—because we are evolution.[1]

### C. *The Dilemma and the Choice*

And now, by the very fact that we have measured the truly cosmic gravity of the sickness that disquiets us, we are put in

[1] There is no such thing as the ' energy of despair ' in spite of what is sometimes said. What those words really mean is a paroxysm of hope against hope. All conscious energy is, like love (and because it is love), founded on hope.

possession of the remedy that can cure it. 'After the long series of transformations leading to man, has the world stopped ? Or, if we are still moving, is it not merely in a circle ?'

The answer to that uneasiness of the modern world springs up by itself when we formulate the dilemma in which the analysis of our action has imprisoned us.

Either nature is closed to our demands for futurity, in which case thought, the fruit of millions of years of effort, is stifled, still-born in a self-abortive and absurd universe. Or else an opening exists—that of the super-soul above our souls ; but in that case the way out, if we are to agree to embark on it, must open out freely onto limitless psychic spaces in a universe to which we can unhesitatingly entrust ourselves.

Between these two alternatives of absolute optimism or absolute pessimism, there is no middle way because by its very nature progress is all or nothing. We are confronted accordingly with two directions and only two : one upwards and the other downwards, and there is no possibility of finding a half-way house.

On neither side is there any tangible evidence to produce. Only, in support of hope, there are rational invitations to an act of faith.

At this cross-roads where we cannot stop and wait because we are pushed forward by life—and obliged to adopt an attitude if we want to go on doing anything whatsoever—what are we going freely to decide ?

To determine man's choice, in his famous wager, Pascal loaded the dice with the lure of boundless gain. Here, when one of the alternatives is weighted with logic and, in a sense, by the promise of a whole world, can we still speak of a simple game of chance ? Have we the right to hesitate ?

The world is too big a concern for that. To bring us into existence it has from the beginning juggled miraculously with too many improbabilities for there to be any risk whatever in committing ourselves further and following it right to the end. If it undertook the task, it is because it can finish it, following the same methods and with the same infallibility with which it began.

In last analysis the best guarantee that a thing should happen is that it appears to us as vitally necessary.

We have said that life, by its very structure, having once been lifted to its stage of thought, cannot go on at all without requiring to ascend ever higher.

This is enough for us to be assured of the two points of which our action has immediate need.

The first is that there is for us, in the future, under some form or another, at least collective, not only survival but also *super-life*.

The second is that, to imagine, discover and reach this superior form of existence, we have only to think and to walk always further in the direction in which the lines passed by evolution take on their maximum coherence.

# BOOK FOUR

# SURVIVAL

CHAPTER ONE

# THE COLLECTIVE ISSUE

## *Preliminary Observation :*
## A Blind Alley to be Avoided : Isolation

WHEN MAN has realised that he carries the world's fortune in himself and that a limitless future stretches before him in which he cannot founder, his first reflex often leads him along the dangerous course of seeking fulfilment in isolation.

In one example of this—flattering to our private egotism—some innate instinct, justified by reflection, inclines us to think that to give ourselves full scope we must break away as far as possible from the crowd of *others*. Is it not in our aloofness from our fellows, or alternatively in their subjection to ourselves, that we will find that ' utmost limit of ourselves ' which is our declared goal ? The study of the past teaches us that, with the onset of reflection, an element partially liberated from phyletic servitudes began to live *for itself*. So is it not in a line continuous with that initial emancipation that further advance must lie? To be more *alone* so as to increase one's being. Like some radiating substance, mankind would in this case culminate in a dust of active, dissociated particles. This doubtless would not mean that a cluster of sparks would be extinguished in darkness, for that would involve the total death whose hypothesis we have just eliminated by our fundamental option. Rather it would involve the hope that, in the long run, some rays, more penetrating or luckier than others, would finish up by finding the path sought from the outset by consciousness, groping for its consummation. Concentration by decentration from the rest ; solitary, and by dint of solitude the elements of the noosphere capable of being saved would find their salvation at the extreme limit of, and by the very excess of, their individualisation.

It is rare around us for extreme individualism to go beyond the bounds of a philosophy of immediate enjoyment and feel the need to come to terms with the profound requirements of action.

Less theoretical and less extreme, but all the more insidious, is another doctrine of ' progress by isolation ' which, at this very moment, is fascinating large sections of mankind—the doctrine of the selection and election of races. Flattering to collective egotism, keener, nobler and more easily aroused than individual egotism, racialism has the virtue in its perspectives of accepting and extending rigorously, just as they occur, the lines of the tree of life. What indeed does the history of the animate world show us but a succession of ramifications, springing up one after the other, one on the top of the other, through the success and domination of a privileged group ? And why should we be exempt from the general rule ? Why should there not be once again between us the struggle for life and the survival of the fittest ; the trial of strength ? The super-man should, like any other stem, be an offshoot from a single bud of mankind.

Isolation of the individual or isolation of the group : here we have two different forms of the same tactics, each seemingly able to produce a plausible justification by pointing to the methods pursued by life for its development right down to us.

We shall be seeing later wherein lies the attraction (or perversity) of these cynical and brutal theories in which, however, a noble passion may also stir. We shall also see why, faced with one or other of these calls to violence, we cannot help sometimes being deeply responsive. They involve a subtle deformation of a great truth.

What matters at the moment is to see clearly that those in both groups deceive themselves, and us too, inasmuch as, ignoring an essential phenomenon—the 'natural confluence of grains of thought'—they disfigure or hide from our eyes the veritable contours of the noosphere and render biologically impossible the formation of a veritable spirit of the earth.

## 1. THE CONFLUENCE OF THOUGHT

### A. *Forced Coalescence*

*a. Coalescence of Elements.* By their very nature, and at every level of complexity, the elements of the world are able to influence and mutually to penetrate each other by their *within*, so as to combine their 'radial energies' in 'bundles'. While no more than conjecturable in atoms and molecules, this psychic interpenetrability grows and becomes directly perceptible in the case of organised beings. Finally in man, in whom the effects of consciousness attain the present maximum found in nature, it reaches a high degree everywhere. It is written all over the social phenomenon and is, of course, felt by us directly. But at the same time, in this case also, it operates only in virtue of the 'tangential energies' of arrangement and thus under certain conditions of spatial juxtaposition.

And here there intervenes a fact, commonplace at first sight, but through which in reality there transpires one of the most fundamental characteristics of the cosmic structure—the roundness of the earth. The geometrical limitation of a star closed, like a gigantic molecule, upon itself. We have already regarded this as a necessary feature at the origin of the first synthesis and polymerisations on the early earth. Implictly, without our having to say so, it has constantly sustained all the differentiations and all the progress of the biosphere. But what are we to say of its function in the noosphere ?

What would have become of humanity if, by some remote chance, it had been free to spread indefinitely on an unlimited surface, that is to say left only to the devices of its internal affinities ? Something unimaginable, certainly something altogether different from the modern world. Perhaps even nothing at all, when we think of the extreme importance of the role played in its development by the forces of compression.

Originally and for centuries there was no serious obstacle to the human waves expanding over the surface of the globe ; probably this is one of the reasons explaining the slowness of their social evolution. Then, from the Neolithic age onwards, these waves began, as we have seen, to recoil upon themselves. All available space being occupied, the occupiers had to pack in tighter. That is how, step by step, through the simple multiplying effect of generations, we have come to constitute, as we do at present, an almost solid mass of hominised substance.

Now, to the degree that—under the effect of this pressure and thanks to their psychic permeability—the human elements infiltrated more and more into each other, their minds (mysterious coincidence) were mutually stimulated by proximity. And as though dilated upon themselves, they each extended little by little the radius of their influence upon this earth which, by the same token, shrank steadily. What in fact do we see happening in the modern paroxysm ? It has been stated over and over again. Through the discovery yesterday of the railway, the motor car and the aeroplane, the physical influence of each man, formerly restricted to a few miles, now extends to hundreds of leagues or more. Better still : thanks to the prodigious biological event represented by the discovery of electro-magnetic waves, each individual finds himself henceforth (actively and passively) simultaneously present, over land and sea, in every corner of the earth.

Thus, not only through the constant increase in the numbers of its members, but also through the continual augmentation of their area of individual activity, mankind—forced to develop as it is in a confined area—has found itself relentlessly subjected to an intense pressure, a self-accentuating pressure, because each advance in it caused a corresponding expansion in each element.

That is one of the first facts to keep in mind, or we shall vitiate our picture of the future of the world.

Undeniably, quite apart from any hypothesis, the external play of cosmic forces, when combined with the nature—so prone to coalesce—of our thinking souls, operates towards a concentration of the energies of consciousness ; and so powerful is this effort that it even succeeds in subjugating the very constructions of phylogenesis—but we shall be coming to that presently.

*b. Coalescence of the Branches.* Twice already—once in developing the theory and once in outlining the historic phases of anthropogenesis—I called attention to the curious property, peculiar to human lines of descent, of coming into contact and mixing with each other, notably by means of their psychic sheath and social institutions. The moment has now come to make a general survey of the phenomenon and discover its ultimate significance.

What at first sight intrigues the naturalist when he tries to *see* the hominids—not merely in themselves, as anthropologists usually do, but in comparison with other animal forms—is the extraordinary elasticity of their zoological group. Outwardly in man, the anatomical differentiation of a primitive type pursues its course as everywhere in evolution. By genetic effects mutations are produced. By climatic and geographical influences, varieties and races come into existence. Somatically speaking, the 'fanning-out' is present continually in formation and perfectly recognisable. Yet the remarkable thing is that its divergent branches no longer succeed in separating. Under conditions of distribution which in any other initial phylum would have led long ago to the break up into different species, the human verticil as it spreads out remains entire, like a gigantic leaf whose veins, however distinct, remain always joined in a common tissue. With man we find indefinite interfecundation on every level, the blending of genes, anastomoses of races in civilisations or political bodies. Zoologically speaking, mankind offers us the unique spectacle of a 'species' capable of achieving something in which all previous species had failed It.

has succeeded, not only in becoming cosmopolitan, but in stretching a single organised membrane over the earth without breaking it.

To what should we attribute this strange condition if not to a reversal, or more exactly a radical perfectioning, of the ways of life by the operation (at last, and only now possible) of a powerful instrument of evolution—the coalescence upon itself of an entire phylum ?

Here again, at the base of the process, lies the exiguity of the earth on which the living stems are forced by their very growth to writhe and intertwine their living branches like serried shoots of ivy. But this external contact was and would always have remained insufficient to reach a point of conjunction without the new ' binder ' conferred on the human biota by the birth of reflection. Until man came, the most life had managed to realise in the matter of association had been to gather socially together on themselves, one by one, the finer extremities of the same phylum. This resulted in essentially mechanical and family groups, created on a purely ' functional ' impulse of construction, defence or propagation, such as the colony, the hive or the ant-heap—all organisms whose power of association is limited to the offspring of one single mother. From man onwards, thanks to the *universal* framework or support provided by thought, free rein is given to the forces of confluence. At the heart of this new *milieu*, the branches themselves of one and the same group succeed in uniting, or rather they become welded together even before they have managed to separate off.

In this way the differentiation of groups in the course of human phylogenesis is maintained up to a certain point, that is to say so far as—by gropingly creating new types—it is a biological condition of discovery and enrichment. After that (or at the same time)—as happens on a sphere where the meridians separate off at one pole only to come together at the other—this divergence gives place to, and becomes subordinate to, a movement of convergence in which races, peoples and nations consolidate one another and complete one another by mutual fecundation.

Anthropologically, ethnically, socially, morally, we understand nothing about man and can make no valid forecasts of his future, so long as we fail to see that, in his case, 'ramification' (in so far as it still persists) works only with the aim —and under higher forms—of agglomeration and convergence. Formation of verticils, selection, struggle for life—henceforward these are secondary functions, subordinate in man to a task of cohesion, a furling back upon itself of a 'bundle' of potential species around the surface of the earth, a completely new mode of phylogenesis.[1]

## B. *Mega-Synthesis*

The coalescence of elements and the coalescence of stems, the spherical geometry of the earth and psychical curvature of the mind harmonising to counterbalance the individual and collective forces of dispersion in the world and to impose unification—there at last we find the spring and secret of hominisation.

But why should there be unification in the world and what purpose does it serve ?

To see the answer to this ultimate question, we have only to put side by side the two equations which have been gradually formulating themselves from the moment we began trying to situate the phenomenon of man in the world.

Evolution= Rise of consciousness,

Rise of consciousness= Union effected.

The general gathering together in which, by correlated actions of the *without* and the *within* of the earth, the totality of thinking units and thinking forces are engaged—the aggregation in a single block of a mankind whose fragments weld together and interpenetrate before our eyes in spite of (indeed in proportion to) their efforts to separate—all this becomes intelligible from top to bottom as soon as we perceive it as the natural culmination of a cosmic processus of organisation which has

[1] This is what I have called elsewhere 'the human Planetisation'.

never varied since those remote ages when our planet was young.

First the molecules of carbon compounds with their thousands of atoms symmetrically grouped ; next the cell which, within a very small volume, contains thousands of molecules linked in a complicated system ; then the metazoa in which the cell is no more than an almost infinitesimal element ; and later the manifold attempts made sporadically by the metazoa to enter into symbiosis and raise themselves to a higher biological condition.

And now, as a germination of planetary dimensions, comes the thinking layer which over its full extent develops and intertwines its fibres, not to confuse and neutralise them but to reinforce them in the living unity of a single tissue.

Really I can see no coherent, and therefore scientific, way of grouping this immense succession of facts but as a gigantic psycho-biological operation, a sort of *mega-synthesis*, the ' super-arrangement ' to which all the thinking elements of the earth find themselves today individually and collectively subject.

Mega-synthesis in the tangential, and therefore and thereby a leap forward of the radial energies along the principal axis of evolution : ever more complexity and thus ever more consciousness. If that is what really happens, what more do we need to convince ourselves of the vital error hidden in the depths of any doctrine of isolation ? The egocentric ideal of a future reserved for those who have managed to attain egoistically the extremity of ' everyone for himself ' is false and against nature. No element could move and grow except with and by all the others with itself.

Also false and against nature is the racial ideal of one branch draining off for itself alone all the sap of the tree and rising over the death of other branches. To reach the sun nothing less is required than the combined growth of the entire foliage.

The outcome of the world, the gates of the future, the entry into the super-human—these are not thrown open to a few of the privileged nor to one chosen people to the exclusion of all others. They will open only to an advance of *all together*, in a direction in which *all together*[1] can join and find completion in a spiritual renovation of the earth, a renovation whose physical degree of reality we must now consider and whose outline we must make clearer.

## 2. THE SPIRIT OF THE EARTH

### A. *Mankind*

Mankind : the idea of mankind was the first image in terms of which, at the very moment that he awoke to the idea of progress, modern man must have tried to reconcile the hopes of an unlimited future with which he could no longer dispense, with the perspective of the inevitability of his own unavoidable individual death. 'Mankind' was at first a vague entity, felt rather than thought out, in which an obscure feeling of perpetual growth was allied to a need for universal fraternity. Mankind was the object of a faith that was often naïve but whose magic, being stronger than all vicissitudes and criticisms, goes on working with persuasive force upon the present-day masses and on the 'intelligentsia' alike. Whether one takes part in the cult or makes fun of it, even today no-one can escape being haunted or even dominated by the idea of mankind.

In the eyes of the 'prophets' of the eighteenth century, the world appeared really as no more than a jumble of confused and loose relationships ; and the divination of a believer was required to feel the beating heart of that sort of embryo. Now, less than two hundred years later, here we are penetrating (though hardly conscious of the fact) into the reality, at any rate the material reality, of what our fathers expected. In the course of a few generations all sorts of economic and cultural links have been forged around us and they are multiplying in geometric progression. Nowadays, over and above the bread which to simple Neolithic man symbolised food, each man demands his

[1] Even if they do so only under the influence of a few, an *élite*.

daily ration of iron, copper and cotton, of electricity, oil and radium, of discoveries, of the cinema and of international news. It is no longer a simple field, however big, but the whole earth which is required to nourish each one of us. If words have any meaning, is this not like some great body which is being born—with its limbs, its nervous system, its perceptive organs, its memory—the body in fact of that great Thing which had to come to fulfil the ambitions aroused in the reflective being by the newly acquired consciousness that he was at one with and responsible to an evolutionary All ?

Indeed, following logically upon our effort to co-ordinate and organise the lines of the world, it is to an outlook recalling the initial intuition of the first philanthropists that our minds constantly return, with the elimination of individualist and racial heresies. No evolutionary future awaits man except in association with all other men. The dreamers of yesterday glimpsed that. And in a sense we see the same thing. But what we are better able to perceive, because we stand on their shoulders, are its cosmic roots, its particular physical substance, and finally the specific nature of this mankind of which they could only have a presentiment—and which we cannot overlook unless we shut our eyes.

*Cosmic roots.* For the earliest humanitarians, man, in uniting with his fellows, was following a natural precept whose origins people hardly bothered to analyse and hence to measure their gravity. In those days, was not nature still treated as a personage or as a poetic metaphor ? What she required of us at a particular time she might have just thought up yesterday and perhaps would no longer want tomorrow. For us, more aware of the dimensions and structural demands of the world, the forces which converge upon us from without or arise from within and drive us ever closer together, are losing any semblance of arbitrariness and any danger of instability.

Mankind was a fragile and even fictitious construction so long as it could only have a limited, plural and disjointed cosmos as a setting ; but it becomes consistent and at the same time probable as soon as it is brought within the compass of a biological space-time and appears as a continuation of the very lines of the universe amongst other realities as vast as itself.

*Physical stuff.* For many of our contemporaries, mankind still remains something unreal, unless materialised in an absurd way. For some it is only an abstract entity or even a mere conventional expression ; for others it becomes a closely-knit organic group in which the social element can be transcribed literally in terms of anatomy and physiology. It appears either as a general idea, a legal entity, or else as a gigantic animal. In both views we find the same inability, by default or by excess, to think the whole correctly. Does not the only way out of this dead-end lie in introducing boldly into our intellectual framework yet another category to serve for the super-individual ? After all, why not ? Geometry, at first constructed on rational conceptions of size, would have remained stationary if it had not in the end accepted 'e', $\pi$, and other incommensurables as being just as complete and intelligible as any whole number. The calculus would never have resolved the problems posed by modern physics if it had not constantly continued to conceive new functions. For identical reasons biology would not be able to generalise itself on the dimensions of the whole of life without introducing into the scale of values that it now needs to deal with certain stages of being which common experience has hitherto been able to ignore—and in particular that of the *collective*. Yes, from now on we envisage, beside and above individual realities, the collective realities that are not reducible to the component element, yet are in their own way as objective as it is. Is it not in this way that I have been inescapably forced to write so as to translate the movements of life into concepts ?

Phyla, layers, branches, etc. . . .

To the eye that has become adjusted to the perspectives of evolution, the directed groups of phyla, layers, branches, etc. become perforce as clear, as physically real, as any isolated object. And in this particular class of dimensions mankind naturally takes its place. But, for it to become representable to us, it is enough that by a mental re-orientation we should reach the point of seeing it directly, exactly as it is, without attempting to put it into terms of anything simpler which we know already.

*Specific nature.* Here, lastly, we pick up the problem again at the point at which the realisation of the confluence of human thoughts had already led us. Being a collective reality, and therefore *sui generis*, mankind can only be understood to the extent that, leaving behind its body of tangible constructions, we try to determine the particular type of conscious synthesis emerging from its laborious and industrious concentration. It is in the last resort only definable as a mind.

Now from this point of view and in the present condition of things, there are two ways, through two stages, in which we can picture the form mankind will assume tomorrow—either (and this is simpler) as a common power and act of knowing and doing, or (and this goes much deeper) as an organic superaggregation of souls. In short : science or unanimity.

### B. *Science*

Taken in the full modern sense of the word, science is the twin sister of mankind. Born together, the two ideas (or two dreams) grew up together to attain an almost religious valuation in the course of the last century. Subsequently they fell together into the same disrepute. But that does not prevent them, when mutually supporting one another as they do, from continuing to represent (in fact more than ever) the ideal forces upon which our imagination falls back whenever it seeks to materialise in terrestrial form its reasons for believing and hoping.

The future of science . . . As a first approximation it is outlined on our horizon as the establishment of an overall and completely coherent perspective of the universe. There was a time when the only part ascribed to knowledge lay in lighting up for our speculative pleasure the objects ready made and given around us. Nowadays, thanks to a philosophy which has given a meaning and a consecration to our thirst to think all things, we can glimpse that unconsciousness is a sort of ontological inferiority or evil, since the world can only fulfil itself in so far as it expresses itself in a systematic and reflective perception. Even (above all, maybe) in mathematics, is not 'discovery' the bringing into existence of something new ? From this point of view, intellectual discovery and synthesis are no longer merely speculation but creation. Therefore, some physical consummation of things is bound up with the explicit perception we make of them. And therefore, they are (at least partially) right[1] who situate the crown of evolution in a supreme act of collective vision obtained by a pan-human effort of investigation and construction.[2]

*Knowledge for its own sake.* But also, and perhaps still more, *knowledge for power.*

Since its birth, science has made its greatest advances when stimulated by some particular problem of life needing a solution ; and its most sublime theories would always have drifted, rootless, on the flood of human thought if they had not been promptly incorporated into some way of mastering the world. Accordingly the march of humanity, as a prolongation of that of all other animate forms, develops indubitably in the direction of a conquest of matter put to the service of mind. *Increased power for increased action.* But, finally and above all, *increased action for increased being.*

Of old, the forerunners of our chemists strove to find the philosophers' stone. Our ambition has grown since then. It is no longer to make gold but life ; and in view of all that has happened in the last fifty years, who would dare to say that this

[1] Is not this one of Brunschvig's ideas?

[2] One might say that, by virtue of human reflection (both individual and collective), evolution, overflowing the physico-chemical organisation of bodies, turns back upon itself and thereby reinforces itself (see note following) with a new organising power vastly concentric to the first—the cognitive organisation of the universe. To think 'the world' (as physics is beginning to realise) is not merely to register it but to confer upon it a form of unity it would otherwise (i.e. without being thought) be without.

is a mere mirage? With our knowledge of hormones we appear to be on the eve of having a hand in the development of our bodies and even of our brains. With the discovery of genes it appears that we shall soon be able to control the mechanism of organic heredity. And with the synthesis of albuminoids imminent, we may well one day be capable of producing what the earth, left to itself, seems no longer able to produce: a new wave of organisms, an artificially provoked neo-life.[1] Immense and prolonged as the universal groping has been since the beginning, many possible combinations have been able to slip through the fingers of chance and have had to await man's calculated efforts in order to appear. Thought might artificially perfect the thinking instrument itself; life might rebound forward under the collective effect of its reflection. The dream upon which human research obscurely feeds is fundamentally that of mastering, beyond all atomic or molecular affinities, the ultimate energy of which all other energies are merely servants; and thus, by grasping the very mainspring of evolution, seizing the tiller of the world.

I salute those who have the courage to admit that their hopes extend that far; they are at the pinnacle of mankind; and I would say to them that there is less difference than people think between research and adoration. But there is a point I would like them to note, one that will lead us gradually to a more complete form of conquest and adoration. However far science pushes its discovery of the 'essential fire' and however capable it becomes some day of remodelling and perfecting the human element, it will always find itself in the end facing the same problem—how to give to each and every element its final value by grouping them in the unity of an organised whole.

[1] It is what I have called '*the human rebound*' of evolution as correlative and conjugated with *Planetisation.*

### C. *Unanimity*

We have used the term mega-synthesis. Within a better understanding of the collective, it seems to me that the word should be understood without attenuation or metaphors when applied to the sum of all human beings. The universe is necessarily homogeneous in its nature and dimensions. Would it still be so if the loops of its spiral lost one jot or tittle of their degree of reality or consistence in ascending ever higher ? The still unnamed Thing which the gradual combination of individuals, peoples and races will bring into existence, must needs be *supra-physical*, not *infra-physical*, if it is to be coherent with the rest. Deeper than the common act in which it expresses itself, more important than the common power of action from which it emerges by a sort of self-birth, lies reality itself, constituted by the living reunion of reflective particles.

And what does that amount to if not (and it is quite credible) that the stuff of the universe, by becoming thinking, has not yet completed its evolutionary cycle, and that we are therefore moving forward towards some new critical point that lies ahead. In spite of its organic links, whose existence has everwhere become apparent to us, the biosphere has so far been no more than a network of divergent lines, free at their extremities. By effect of reflection and the recoils it involves, the loose ends have been tied up, and the noosphere tends to constitute a single closed system in which each element sees, feels, desires and suffers for itself the same things as all the others at the same time.

We are faced with a harmonised collectivity of consciousnesses equivalent to a sort of super-consciousness. The idea is that of the earth not only becoming covered by myriads of grains of thought, but becoming enclosed in a single thinking envelope so as to form, functionally, no more than a single vast grain of thought on the sidereal scale, the plurality of individual reflections grouping themselves together and reinforcing one another in the act of a single unanimous reflection.

This is the general form in which, by analogy and in symmetry with the past, we are led scientifically to envisage the future of mankind, without whom no terrestrial issue is open to the terrestrial demands of our action.

To the common sense of the ' man in the street ' and even to a certain philosophy of the world to which nothing is possible save what has always been, perspectives such as these will seem highly improbable. But to a mind become familiar with the fantastic dimensions of the universe they will, on the contrary, seem quite natural, because they are simply proportionate with the astronomical immensities.

In the direction of thought, could the universe terminate with anything less than the measureless—any more than it could in the direction of time and space ?

One thing at any rate is sure—from the moment we adopt a thoroughly realistic view of the noosphere and of the hyper-organic nature of social bonds, the present situation of the world becomes clearer ; for we find a very simple meaning for the profound troubles which disturb the layer of mankind at this moment.

The two-fold crisis whose onset began in earnest as early as the Neolithic age and which rose to a climax in the modern world, derives in the first place from a *mass-formation* (we might call it a ' planetisation ') of mankind. Peoples and civilisations reached such a degree either of frontier contact or economic interdependence or psychic communion that they could no longer develop save by interpenetration of one another. But it also arises out of the fact that, under the combined influence of machinery and the super-heating of thought, we are witnessing *a formidable upsurge of unused powers.* Modern man no longer knows what to do with the time and the potentialities he has unleashed. We groan under the burden of this wealth. We are haunted by the fear of ' unemployment '. Sometimes we are tempted to trample this super-abundance back into the matter from which it sprang without stopping to think how impossible and monstrous such an act against nature would be.

When we consider the increasing compression of elements at the heart of a free energy which is also relentlessly increasing, how can we fail to see in this two-fold phenomenon the two perennial symptoms of a leap forward of the ' radial '—that is to say, of a new step in the genesis of mind ?

In order to avoid disturbing our habits we seek in vain to settle international disputes by adjustments of frontiers—or we treat as ' leisure ' (to be whiled away) the activities at the disposal of mankind. As things are now going it will not be long before we run full tilt into one another. Something will explode if we persist in trying to squeeze into our old tumble-down huts the material and spiritual forces that are henceforward on the scale of a world.

A new domain of psychical expansion—that is what we lack. And it is staring us in the face if we would only raise our heads to look at it.

Peace through conquest, work in joy. These are waiting for us beyond the line where empires are set up against other empires, in an interior totalisation of the world upon itself, in the unanimous construction of a *spirit of the earth.*

How is it then that our first efforts towards this great goal seem merely to take us farther from it ?

CHAPTER TWO

# BEYOND THE COLLECTIVE: THE HYPER-PERSONAL

## *Another Preliminary Observation*

### A Feeling to be overcome : Discouragement

THE REASONS behind the scepticism regarding mankind which is fashionable among 'enlightened' people today are not merely of a representative order. Even when the intellectual difficulties of the mind in conceiving the collective and visualising space-time have been overcome, we are left with another and perhaps a still more serious form of hesitation which is bound up with the incoherent aspect presented by the world of men today. The nineteenth century had lived in sight of a promised land. It thought that we were on the threshold of a Golden Age, lit up and organised by science, warmed by fraternity. Instead of that, we find ourselves slipped back into a world of spreading and ever more tragic dissension. Though possible and even perhaps probable in theory, the idea of a spirit of the earth does not stand up to the test of experience. No, man will never succeed in going beyond man by uniting with himself. That Utopia must be abandoned as soon as possible and there is no more to be said.

To explain or efface the appearances of a setback which, if it were true, would not only dispel a beautiful dream but encourage us to weigh up a radical absurdity of the universe, I would like to point out in the first place that to speak of experience—of the results of experience—in such a connection is premature to say the least of it. After all half a million years, perhaps even a million, were required for life to pass from the pre-hominids to modern man. Should we now start wringing our hands because, less than two centuries after glimpsing a higher state, modern man is still at loggerheads with himself ? Once again we have got things out of focus. To have understood the immensity around us, behind us, and in front of us is already a first step. But if to this perception of depth another perception, that of *slowness*, be not added, we must realise that the transposition of values remains incomplete and that it can beget for our gaze nothing but an impossible world. Each dimension has its proper rhythm. Planetary movement involves planetary majesty. Would not humanity seem to us altogether static if, behind its history, there were not the endless stretch of its pre-history ? Similarly, and despite an almost explosive acceleration of noogenesis at our level, we cannot expect to see the earth transform itself under our eyes in the space of a generation. Let us keep calm and take heart.

In spite of all evidence to the contrary, mankind may very well be advancing all round us at the moment—there are in fact many signs whereby we can reasonably suppose that it is advancing. But, if it is doing so, it must be—as is the way with very big things—doing so almost imperceptibly.

This point is of the utmost importance and must never be lost sight of. To have made it does not, however, allay the most acute of our fears. After all we need not mind very much if the light on the horizon appears stationary. What does matter is when it seems to be going out. If only we could believe that we were merely motionless ! But does it not sometimes seem that we are actually being blocked in our advance or even swallowed up from behind—as though we were in the grip of some ineluctable forces of mutual repulsion and materialisation. *Repulsion.* I have spoken of the formidable pressures which hem in the human particles in the present-day world, both individuals and peoples being forced in an extreme way, geographically and psychologically, up against one another. Now the strange fact is that, in spite of the strength of these energies bringing men together, thinking units do not seem capable of falling within their radius of internal attraction. Leaving aside individual cases, where sexual forces or some extraordinary and transitory common passion come into play, men are hostile or at least closed to one another. Like a powder whose particles, however compressed, refuse to enter into molecular contact, deep down men exclude and repel one another with all their might : unless (and this is worse still) their mass forms in such a way that, instead of the expected *mind*, a new wave of determinism surges up—that is to say, of materiality.

*Materialisation.* Here I am not only thinking of the laws of large numbers which, irrespective of their secret ends, enslave by structure each newly-formed multitude. As with every other form of life, man, to become fully man, had to become legion. And, before becoming organised, a legion is necessarily prey to the play, however directed it be, of chance and probability. There are imponderable currents which, from fashion and rates of exchange to political and social revolutions, make us all the slaves of the obscure seethings of the human mass. However spiritualised we suppose its elements to be, every aggregate of consciousness, so long as it is not harmonised, envelops itself automatically (at its own level) with a veil of 'neo-matter', superimposed upon all other forms of matter—matter, the 'tangential' aspect of every living mass in course of unification. Of course we must react to such conditions ; but with the satisfaction of knowing that they are only the sign of and price paid for progress. But what are we to say of the other slavery, the one which gains ground in the world in very proportion to the efforts we make to organise ourselves ?

At no previous period of history has mankind been so well equipped nor made such efforts to reduce its multitudes to order. We have 'mass movements'—no longer the hordes streaming down from the forests of the north or the steppes of Asia, but 'the Million' scientifically assembled. The Million in rank and file on the parade ground ; the Million standardised in the factory; the Million motorised—and all this only ending up with Communism and National-Socialism and the most ghastly fetters. So we get the crystal instead of the cell; the ant-hill instead of brotherhood. Instead of the upsurge of consciousness which we expected, it is mechanisation that seems to emerge inevitably from totalisation.

*'Eppur si muove!'*

In the presence of such a profound perversion of the rules of noogenesis, I hold that our reaction should be not one of despair but of a determination to re-examine ourselves. When an energy runs amok, the engineer, far from questioning the power itself, simply works out his calculations afresh to see how it can be brought better under control. Monstrous as it is, is not modern totalitarianism really the distortion of something magnificent, and thus quite near to the truth? There can be no doubt of it: the great human machine is designed to work and *must* work—by producing a super-abundance of mind. If it does not work, or rather if it produces only matter, this means that it has gone into reverse.

Is it not possible that in our theories and in our acts we have neglected to give due place to the person and the forces of *personalisation*?

## 1. THE CONVERGENCE OF THE PERSON AND THE OMEGA POINT

### A. *The Personal Universe*

Unlike the primitives who gave a face to every moving thing, or the early Greeks who defined all the aspects and forces of nature, modern man is obsessed by the need to depersonalise (or impersonalise) all that he most admires. There are two reasons for this tendency, The first is *analysis*, that marvellous instrument of scientific research to which we owe all our advances but which, breaking down synthesis after synthesis, allows on soul after another to escape, leaving us confronted with a pile of dismantled machinery, and evanescent particles. The second reason lies in the discovery of the sidereal world, so vast that it seems to do away with all proportion between our own being and the dimensions of the cosmos around us. Only one reality seems to survive and be capable of succeeding and spanning the infinitesimal and the immense: energy—that floating, universal entity from which all emerges and into which all falls back as into an ocean ; energy, the new spirit ; the new god. So, at the world's Omega, as at its Alpha, lies the Impersonal.

Under the influence of such impressions as these, it looks as though we have lost both respect for the person and understanding of his true nature. We end up by admitting that to be pivoted on oneself, to be able to say 'I', is the privilege (or rather the blemish) of the element in the measure to which the latter closes the door on all the rest and succeeds in setting himself up at the antipodes of the All. In the opposite direction we conceive the 'ego' to be diminishing and eliminating itself, with the trend to what is most real and most lasting in the world, namely the Collective and the Universal. Personality is seen as a specifically corpuscular and ephemeral property ; a prison from which we must try to escape.

Intellectually, that is more or less where we stand today.

Yet if we try, as I have done in this essay, to pursue the logic and coherence of facts to the very end, we seem to be led to the precisely opposite view by the notions of space-time and evolution.

We have seen and admitted that evolution is an ascent towards consciousness. That is no longer contested even by the most materialistic, or at all events by the most agnostic of humanitarians. Therefore it should culminate forwards in some sort of supreme consciousness. But must not that consciousness, if it is to be supreme, contain in the highest degree what is the perfection of our consciousness—the illuminating involution of the being upon itself ? It would manifestly be an error to extend the curve of hominisation in the direction of a state of diffusion.

It is only in the direction of hyper-reflection—that is to say, hyper-personalisation—that thought can extrapolate itself. Otherwise how could it garner our conquests which are all made in the field of what is reflected ? At first sight we are disconcerted by the association of an Ego with what is the All. The utter disproportion of the two terms seems flagrant, almost laughable. That is because we have not sufficiently meditated upon the three-fold property possessed by every consciousness : (i) of centring *everything* partially upon itself ; (ii) of being able to centre itself upon itself *constantly* ; and (iii) of being brought *more* by this very super-centration *into association with all the other centres* surrounding it. Are we not at every instant living the experience of a universe whose immensity, by the play of our senses and our reason, is gathered up more and more simply in each one of us ? And in the establishment now proceeding through science and the philosophies of a collective human *Weltanschauung* in which every one of us co-operates and participates, are we not experiencing the first symptoms of an aggregation of a still higher order, the birth of some single centre from the convergent beams of millions of elementary centres dispersed over the surface of the thinking earth ?

All our difficulties and repulsions as regards the opposition between the All and the Person would be dissipated if only we understood that, by structure, the noosphere (and more generally the world) represent a whole that is not only closed but also *centred*. Because it contains and engenders consciousness, space-time is necessarily *of a convergent nature*. Accordingly its enormous layers, followed in the right direction, must somewhere ahead become involuted to a point which we might call *Omega*, which fuses and consumes them integrally in itself. However immense the sphere of the world may be, it only exists and is finally perceptible in the directions in which its radii meet—even if this were beyond time and space altogether. Better still : the more immense this sphere, the richer and deeper and hence the more conscious is the point at which the 'volume of being' that it embraces is concentrated ; because the mind, seen from our side, is essentially the power of synthesis and organisation.

Seen from this point of view, the universe, without losing any of its immensity and thus without suffering any anthropomorphism, begins to take shape : since to think it, undergo it and make it act, it is *beyond* our souls that we must look, *not the other way round.* In the perspective of a noogenesis, time and space become truly humanised—or rather super-humanised. Far from being mutually exclusive, the Universal and Personal (that is to say, the 'centred') grow in the same direction and culminate simultaneously in each other.

It is therefore a mistake to look for the extension of our being or of the noosphere in the Impersonal. The Future-Universal could not be anything else but the Hyper-Personal—at the Omega Point.

### B. *The Personalising Universe*

*Personalisation.* It is by this internal deepening of consciousness upon itself that we have characterised (Book III, Chapter I, Section I) the particular destiny of the element that has become fully itself by crossing the threshold of reflection—and there, a regards the fate of individual human beings—we brought our inquiry to a provisional halt. *Personalisation* : the same type of progress reappears here, but this time it defines the collective future of totalised grains of thought. There is an identical function for the element as for the sum of the elements brought together in a synthesis. How can we conceive and foresee that the two movements harmonise ? How, without being impeded or deformed, can the innumerable particular curves be inscribed or even prolonged in their common envelope ?

The time has come to tackle this problem, and, for that purpose, to analyse still further the nature of the personal centre of convergence upon whose existence hangs the evolutionary equilibrium of the noosphere. What should this higher pole of evolution be, in order to fulfil its role ?

It is by definition in Omega that—in its flower and its integrity—the hoard of consciouness liberated little by little on earth by noogenesis adds itself together and concentrates. So much has already been accepted. But what exactly do we mean, what is implied, when we use the apparently simple phrase 'addition of consciousness'?

When we listen to the disciples of Marx, we might think it was enough for mankind (for its growth and to justify the sacrifices imposed on us) to gather together the successive acquisitions we bequeath to it in dying—our ideas, our discoveries, our works of art, our example. Surely this imperishable treasure is the best part of our being.

Let us reflect a moment, and we shall soon see that for a universe which, by hypothesis, we admitted to be a 'collector and custodian of consciousness', the mere hoarding of these remains would be nothing but a colossal wastage. What passes from each of us into the mass of humanity by means of invention, education and diffusion of all sorts is admittedly of vital importance. I have sufficiently tried to stress,its phyletic value and no one can accuse me of belittling it. But with that accepted, I am bound to admit that, in these contributions to the collectivity, far from transmitting the most precious, we are bequeathing, at the utmost, only the shadow of ourselves. Our works? But even in the interest of life in general, what is the work of works for man if not to establish, in and by each one of us, an absolutely original centre in which the universe reflects itself in a unique and inimitable way? And those centres are our very selves and personalities. The very centre of our consciousness, deeper than all its radii; that is the essence which Omega, if it is to be truly Omega, must reclaim. And this essence is obviously not something of which we can dispossess ourselves for the benefit of others as we might give away a coat or pass on a torch. For we are the very flame of that torch. To communicate itself, my ego must subsist through abandoning itself or the gift will fade away. The conclusion is inevitable that the concentration of a conscious universe would be unthinkable if it did not reassemble in itself *all consciousnesses* as well as all *the conscious* ; each particular consciousness remaining conscious of itself at the end of the operation, and even (this must absolutely be understood) each particular consciousness becoming still more itself and thus more clearly distinct from others the closer it gets to them in Omega.

The exaltation, not merely the conservation, of elements by convergence : what, after all, could be more simple, and more thoroughly in keeping with all we know ?

In any domain—whether it be the cells of a body, the members of a society or the elements of a spiritual synthesis—*union differentiates*. In every organised whole, the parts perfect themselves and fulfil themselves. Through neglect of this universal rule many a system of pantheism has led us astray to the cult of a great All in which individuals were supposed to be merged like a drop in the ocean or like a dissolving grain of salt. Applied to the case of the summation of consciousnesses, the law of union rids us of this perilous and recurrent illusion. No, following the confluent orbits of their centres, the grains of consciousness do not tend to lose their outlines and blend, but, on the contrary, to accentuate the depth and incommunicability of their *egos*. The more ' other ' they become in conjunction, the more they find themselves as ' self '. How could it be otherwise since they are steeped in Omega? Could a centre dissolve? Or rather, would not its particular way of dissolving be to supercentralise itself ?

Thus, under the influence of these two factors—the essential immiscibility of consciousnesses, and the natural mechanism of all unification—the only fashion in which we could correctly express the final state of a world undergoing psychical concentration would be as a system whose unity coincides with a paroxysm of harmonised complexity. Thus it would be mistaken to represent Omega to ourselves simply as a centre born of the fusion of elements which it collects, or annihilating them in itself. By its structure Omega, in its ultimate principle, can only be a *distinct Centre radiating at the core of a system of centres* ; a grouping in which personalisation of the All and personalisations of the elements reach their maximum, simultaneously and without merging, under the influence of a supremely autonomous focus of union.[1] That is the only picture which emerges when we try to apply the notion of collectivity with remorseless logic to a granular whole of thoughts.

And at this point we begin to see the motives for the fervour and the impotence which accompany every egoistic solution of life. Egoism, whether personal or racial, is quite rightly excited by the idea of the element ascending through faithfulness to life, to the extremes of the incommunicable and the exclusive that it holds within it. It *feels* right. Its only mistake, but a fatal one, is *to confuse individuality with personality.* In trying to separate itself as much as possible from others, the element individualises itself ; but in doing so it becomes retrograde and seeks to drag the world backwards towards plurality and into matter. In fact it diminishes itself and loses itself. To be fully ourselves it is in the opposite direction, in the direction of convergence with all the rest, that we must advance—towards the ' other '. The peak of ourselves, the acme of our originality, is not our individuality but our person ; and according to the evolutionary structure of the world, we can only find our person by uniting together. There is no mind without synthesis. The same law holds good from top to bottom. The true ego grows in inverse proportion to ' egoism '. Like the Omega which attracts it, the element only becomes personal when it universalises itself.[2]

There is, however, an obvious and essential proviso to be made. For the human particles to become really personalised under the creative influence of union—according to the preceding analysis—not every kind of union will do. Since it is a question of achieving a synthesis of centres, it is centre to centre that they

[1] It is for this central focus, necessarily autonomous, that we shall henceforward reserve the expression ' Omega Point '.

[2] Conversely, it only universalises itself properly in becoming super-personal. There is all the difference (and ambiguity) between the true and the false political or religious mysticisms. By the latter man is destroyed ; by the former he is fulfilled by ' becoming lost in the greater than himself '.

must make contact and *not otherwise*. Thus, amongst the various forms of psychic inter-activity animating the noosphere, the energies we must identify, harness and develop before all others are those of an 'intercentric' nature, if we want to give effective help to the progress of evolution in ourselves.

Which brings us to the problem of love.

## 2. LOVE AS ENERGY

We are accustomed to consider (and with what a refinement of analysis!) only the sentimental face of love, the joy and miseries it causes us. It is in its natural dynamism and its evolutionary significance that I shall be dealing with it here, with a view to determining the ultimate phases of the phenomenon of man.

Considered in its full biological reality, love—that is to say, the affinity of being with being—is not peculiar to man. It is a general property of all life and as such it embraces, in its varieties and degrees, all the forms successively adopted by organised matter. In the mammals, so close to ourselves, it is easily recognised in its different modalities: sexual passion, parental instinct, social solidarity, etc. Farther off, that is to say lower down on the tree of life, analogies are more obscure until they become so faint as to be imperceptible. But this is the place to repeat what I said earlier when we were discussing the '*within* of things'. If there were no real internal propensity to unite, even at a prodigiously rudimentary level—indeed in the molecule itself—it would be physically impossible for love to appear higher up, with us, in 'hominised' form. By rights, to be certain of its presence in ourselves, we should assume its presence, at least in an inchoate form, in everything that is. And in fact if we look around us at the confluent ascent of consciousnesses, we see it is not lacking anywhere. Plato felt this and has immortalised the idea in his *Dialogues*. Later, with thinkers like Nicolas of Cusa, mediaeval philosphy returned technically to the same notion. Driven by the forces of love, the fragments of the world seek each other so that the world may come to being. This is no metaphor; and it is much more than poetry. Whether as a force or a curvature, the universal gravity of bodies, so striking to us, is merely the reverse or shadow of that which really moves nature. To perceive cosmic energy 'at the fount' we must, if there is a *within* of things, go down into the internal or radial zone of spiritual attractions.

Love in all its subtleties is nothing more, and nothing less, than the more or less direct trace marked on the heart of the element by the psychical convergence of the universe upon itself.

This, if I am not mistaken is the ray of light which will help us to see more clearly around us.

We are distressed and pained when we see modern attempts at human collectivisation ending up, contrary to our expectations and theoretical predictions, in a lowering and an enslavement of consciousnesses. But so far how have we gone about the business of unification? A material situation to be defended; a new industrial field to be opened up, better conditions for a social class or less favoured nations—those are the only and very mediocre grounds on which we have so far tried to get together. There is no cause to be surprised if, in the footsteps of animal societies, we become mechanised in the very play of association. Even in the supremely intellectual activity of science (at any rate as long as it remains purely speculative and abstract) the impact of our souls only operates obliquely and indirectly. Contact is still superficial, involving the danger of yet another servitude. Love alone is capable of uniting living beings in such a way as to complete and fulfil them, for it alone takes them and joins them by what is deepest in themselves. This is a fact of daily experience. At what moment do lovers come into the most complete possession of themselves if not when they say they are lost in each other? In truth, does not love every instant achieve all around us, in the couple or the team, the magic feat, the feat reputed to be contradictory, of 'personalising' by totalising? And if that is what it can achieve daily on a small scale, why should it not repeat this one day on world-wide dimensions?

Mankind, the spirit of the earth, the synthesis of individuals and peoples, the paradoxical conciliation of the element with the whole, and of unity with multitude—all these are called Utopian and yet they are biologically necessary. And for them to be incarnated in the world all we may well need is to imagine our power of loving developing until it embraces the total of men and of the earth.

It may be said that this is the precise point at which we are invoking the impossible. Man's capacity, it may seem, is confined to giving his affection to one human being or to very few. Beyond that radius the heart does not carry, and there is only room for cold justice and cold reason. To love all and everyone is a contradictory and false gesture which only leads in the end to loving no-one.

To that I would answer that if, as you claim, a universal love is impossible, how can we account for that irresistible instinct in our hearts which leads us towards unity whenever and in whatever direction our passions are stirred ? A sense of the universe, a sense of the *all*, the nostalgia which seizes us when confronted by nature, beauty, music—these seem to be an expectation and awareness of a Great Presence. The 'mystics' and their commentators apart, how has psychology been able so consistently to ignore this fundamental vibration whose ring can be heard by every practised ear at the basis, or rather at the summit, of every great emotion ? Resonance to the All—the keynote of pure poetry and pure religion. Once again : what does this phenomenon, which is born with thought and grows with it, reveal if not a deep accord between two realities which seek each other ; the severed particle which trembles at the approach of 'the rest' ?

We are often inclined to think that we have exhausted the various natural forms of love with a man's love for his wife, his children, his friends and to a certain extent for his country. Yet precisely the most fundamental form of passion is missing from this list, the one which, under the pressure of an involuting universe, precipitates the elements one upon the other in the

Whole—cosmic affinity and hence cosmic sense. A universal love is not only psychologically possible ; it is the only complete and final way in which we are able to love.

But, with this point made, how are we to explain the appearance all around us of mounting repulsion and hatred ? If such a strong potentiality is besieging us from within and urging us to union, what is it waiting for to pass from potentiality to action ? Just this, no doubt : that we should overcome the 'anti-personalist' complex which paralyses us, and make up our minds to accept the possibility, indeed the reality, of some *source* of love and *object* of love at the summit of the world above our heads. So long as it absorbs or appears to absorb the person, collectivity kills the love that is trying to come to birth. As such collectivity is essentially unlovable. That is where philanthropic systems break down. Common sense is right. It is impossible to give oneself to an anonymous number. But if the universe ahead of us assumes a face and a heart, and so to speak personifies itself,[1] then in the atmosphere created by this focus the elemental attraction will immediately blossom. Then, no doubt, under the heightened pressure of an infolding world, the formidable energies of attraction, still dormant between human molecules, will burst forth.

The discoveries of the last hundred years, with their unitary perspectives, have brought a new and decisive impetus to our sense of the world, to our sense of the earth, and to our human sense. Hence the rise of modern pantheism. But this impetus will only end by plunging us back into super-matter unless it leads us towards someone.

For the failure that threatens us to be turned into success, for the concurrence of human monads to come about, it is necessary and sufficient for us that we should extend our science to its farthest limits and recognise and accept (as being necessary to close and balance space-time) not only some vague future

[1] Not, of course, by becoming a person, but by charging itself at the very heart of its development with the dominating and unifying influence of a focus of personal energies and attractions.

existence, but also, as I must now stress, the radiation *as a present reality* of that mysterious centre of our centres which I have called Omega.

## 3. THE ATTRIBUTES OF THE OMEGA POINT

After allowing itself to be captivated in excess by the charms of analysis to the extent of falling into illusion, modern thought is at last getting used once more to the idea of the creative value of synthesis in evolution. It is beginning to see that there is definitely *more* in the molecule than in the atom, *more* in the cell than in the molecule, *more* in society than in the individual, and *more* in mathematical construction than in calculations and theorems. We are now inclined to admit that at each further degree of combination *something* which is irreducible to isolated elements *emerges* in a new order. And with this admission, consciousness, life and thought are on the threshold of acquiring a right to existence in terms of science. But science is nevertheless still far from recognising that this *something* has a particular value of independence and solidity. For, born of an incredible concourse of chances on a precariously assembled edifice, and failing to create any measurable increase of energy by their advent, are not these ' creatures of synthesis ', from the experimental point of view, the most beautiful as well as the most fragile of things ? How could they anticipate or survive the ephemeral union of particles on which their souls have alighted ? So in the end, in spite of a half-hearted conversion to spiritual views, it is still on the *elementary* side—that is, towards matter infinitely diluted —that physics and biology look to find the eternal and the Great Stability.

In conformity with this state of mind the idea that some Soul of souls should be developing at the summit of the world is not as strange as might be thought from the present-day views of human reason. After all, is there any other way in which our thought can generalise the Principle of Emergence ?[1] At the same time, as this Soul coincides with a supremely improbable coincidence of the totality of elements and causes, it remains understood or implied that it could not form itself save at an extremely distant future and in a total dependence on the reversible laws of energy.

Yet it is precisely from these two restrictions (fragility and distance), both incompatible to my mind with the nature and function of Omega, that we want to rid ourselves—and this for two positive reasons, one of love, the other of survival.

First of all the *reason of Love*. Expressed in terms of internal energy, the cosmic function of Omega consists in initiating and maintaining within its radius the unanimity of the world's 'reflective' particles. But how could it exercise this action were it not in some sort loving and lovable *at this very moment* ? Love, I said, dies in contact with the impersonal and the anonymous. With equal infallibility it becomes impoverished with remoteness in space—and still more, much more, with difference in time. For love to be possible there must be co-existence. Accordingly, however marvellous its foreseen figure, Omega could never even so much as equilibrate the play of human attractions and repulsions if it did not act with equal force, that is to say with the same stuff of proximity. With love, as with every other sort of energy, it is within the existing datum that the lines of force must at every instant come together. Neither an ideal centre, nor a potential centre could possibly suffice. A present and real noosphere goes with a real and present centre. To be supremely attractive, Omega must be supremely present.

In addition, the *reason of survival*. To ward off the threat of disappearance, incompatible with the mechanism of reflective activity, man tries to bring together in an ever vaster and more permanent subject the collective principle of his acquisitions—civilisation, humanity, the spirit of the earth. Associated in these enormous entities, with their incredibly slow rhythm of evolu-

[1] See the quotation from J. B. S. Haldane in footnote p. 57.

tion, he has the impression of having escaped from the destructive action of time.[1]

But by doing this he has only pushed back the problem. For after all, however large the radius traced within time and space, does the circle ever embrace anything but the perishable ? So long as our constructions rest with all their weight on the earth, they will vanish with the earth. The radical defect in all forms of belief in progress, as they are expressed in positivist credos, is that they do not definitely eliminate death. What is the use of detecting a focus of any sort in the van of evolution if that focus can and must one day disintegrate? To satisfy the ultimate requirements of our action, Omega must be independent of the collapse of the forces with which evolution is woven.

Actuality, irreversibility. There is only one way in which our minds can integrate into a coherent picture of noogenesis these two essential properties of the autonomous centre of all centres, and that is to resume and complement our Principle of Emergence. In the light of our experience it is abundantly clear that emergence *in the course of evolution* can only happen successively and with mechanical dependence on what precedes it. First the grouping of the elements ; then the manifestation of 'soul' whose operation only betrays, from the point of view of energy, a more and more complex and sublimated involution of the powers transmitted by the chains of elements. The radial function of the tangential : a pyramid whose apex is supported from below : that is what we see during the course of the process. And it is in the very same way that Omega itself is discovered to us at the end of the whole processus, inasmuch as in it the movement of synthesis culminates. Yet we must be careful to note that under this evolutive facet Omega still only reveals *half of itself*. While being the last term of its series, it is also *outside all series*. Not only does it crown, but it closes. Otherwise the sum would fall short of itself, in organic contradiction with the whole operation. When, going beyond the elements, we

[1] See for example that curious book by Wells, *The Anatomy of Frustration*, which eloquently bears witness to the faith and the misgivings of modern man.

come to speak of the conscious Pole of the world, it is not enough to say that it *emerges* from the rise of consciousnesses: we must add that from this genesis it has already *emerged*; without which it could neither subjugate into love nor fix in incorruptibility. If by its very nature it did not escape from the time and space which it gathers together, it would not be Omega.

Autonomy, actuality, irreversibility, and thus finally transcendence are the four attributes of Omega. In this way we round off without difficulty the scheme left incomplete at the end of our second chapter, where we sought to enclose the energy-complex of our universe.

In Omega we have in the first place the principle we needed to explain both the persistent march of things towards greater consciousness, and the paradoxical solidity of what is most fragile. Contrary to the appearances still admitted by physics, the Great Stability is not at the bottom in the infra-elementary sphere, but at the top in the ultra-synthetic sphere. It is thus entirely by its tangential envelope that the world goes on dissipating itself in a chance way into matter. By its radial nucleus it finds its shape and its natural consistency in gravitating against the tide of probability towards a divine focus of mind which draws it onward.

Thus something in the cosmos escapes from entropy, and does so more and more.

During immense periods in the course of evolution, the radial, obscurely stirred up by the action of the *Prime Mover ahead*, was only able to express itself, in diffuse aggregates, in animal consciousness. And at that stage, not being able, above them, to attach themselves to a support whose order of simplicity was greater than their own, the nuclei were hardly formed before they began to disaggregate. But as soon as, through reflection, a type of unity appeared no longer closed or even centred, but punctiform, the sublime physics of centres came into play. When they became centres, and therefore persons, the elements could at last begin to react, directly as such, to the personalising action of the centre of centres. When consciousness broke through the critical surface of hominisation, it really passed from divergence to convergence and changed, so to speak, both hemisphere and pole. Below that critical 'equator' lay the relapse into multiplicity; above it, the plunge into growing and irreversible unification. Once formed, a reflective centre can no longer change except by involution upon itself. To outward appearance, admittedly, man disintegrated just like any animal. But here and there we find an inverse function of the phenomenon. By death, in the animal, the radial is reabsorbed into the tangential, while in man it escapes and is liberated from it. It escapes from entropy by turning back to Omega: the *hominisation* of death itself.

Thus from the grains of thought forming the veritable and indestructable atoms of its stuff, the universe—a well-defined universe in the outcome—goes on building itself above our heads in the inverse direction of matter which vanishes. The universe is a collector and conservator, not of mechanical energy, as we supposed, but of persons. All round us, one by one, like a continual exhalation, 'souls' break away, carrying upwards their incommunicable load of consciousness. One by one, yet not in isolation. Since, for each of them, by the very nature of Omega, there can only be one possible point of definitive emersion—that point at which, under the synthesising action of personalising union, the noosphere (furling its elements upon themselves as it too furls upon itself) will reach collectively its point of convergence—at the 'end of the world'.

CHAPTER THREE

# THE ULTIMATE EARTH

WE HAVE seen that without the involution of matter upon itself, that is to say, without the closed chemistry of molecules, cells and phyletic branches, there would never have been either biosphere or noosphere. In their advent and their development, life and thought are not only accidentally, but also structurally, bound up with the contours and destiny of the terrestrial mass.

But, on the other hand, we now see ahead of us a psychical centre of universal drift, transcending time and space and thus essentially extra-planetary, to sustain and equilibrate the surge of consciousnesses.

The idea is that of noogenesis ascending irreversibly towards Omega through the strictly limited cycle of a geogenesis. At a given moment in the future, under some influence exerted by one or the other of these curves or of both together, it is inevitable that the two branches should separate. However convergent it be, evolution cannot attain to fulfilment on earth except through a point of dissociation.

With this we are introduced to a fantastic and inevitable event which now begins to take shape in our perspective, the event which comes nearer with every day that passes : the end of all life on our globe, the death of the planet, the ultimate phase of the phenomenon of man.

No one would dare to picture to himself what the noosphere will be like in its final guise, no one, that is, who has glimpsed however faintly the incredible potential of unexpectedness accumulated in the spirit of the earth. The end of the world defies imagination. But if it would be absurd to try to describe it, we may none the less—by making use of the lines of approach already laid down—to some extent foresee the significance and circumscribe the forms.

What the ultimate earth cannot be in a universe of conscious substance; how will it take shape; and what it will probably be—those are the questions I want to raise, coldly and logically, in no way apocalyptically, not so much for the sake of affirming anything as to give food for thought.

## 1. PROGNOSTICS TO BE SET ASIDE

When the end of the world is mentioned, the idea that leaps into our minds is always one of catastrophe.

Generally we think of a sidereal cataclysm. There are so many stars hurtling around and brushing past ; there are those exploding worlds on the horizon ; so, surely, by the implacable laws of chance, our turn will come sooner or later and we shall be stricken and killed ; or, at the least, we shall have to face a slow death in our prison.

Since physics has discovered that all energy runs down, we seem to feel the world getting a shade chillier every day. That cooling-off to which we were condemned has been partially compensated for by another discovery, that of radio-activity, which has happily intervened to compensate and delay the imminent cooling. The astronomers are now in a position to guarantee that, if all goes as it should, we have at any rate several hundred million years ahead of us. So we can breathe again. Yet, though the settlement is postponed, the shadow grows longer.

And will mankind still be there to watch the evening fall ? In the interim, apart from the cosmic mishaps that lie in wait for us, what will happen in the living layer of the earth ? With age and increasing complication, we are ever more threatened by internal dangers at the core of both the biosphere and the noosphere. Onslaughts of microbes, organic counter-evolutions, sterility, war, revolution—there are so many ways of coming to an end. Yet perhaps anything would be better than a long-drawn-out senility.

We are well aware of these different eventualities. We have turned them over in our minds. We have read descriptions of them in the novels of the Goncourts, Benson and Wells, or in scientific works signed by famous names. Each one of them is perfectly feasible. We could very well, and at any moment, be crushed by a gigantic comet. And, equally true, tomorrow the earth might quake and collapse under our feet. Taken individually, each human will can repudiate the task of ascending higher towards union. And yet, on the strength of all we learn from past evolution, I feel entitled to say that we have nothing whatever to fear from these manifold disasters *in so far as* they imply the idea of premature accident or failure. However possible they may be in theory, we have higher reasons for being sure *that they will not happen.*

All pessimistic representations of the earth's last days—whether in terms of cosmic catastrophe, biological disruptions or simply arrested growth or senility—have this in common: that they take the characteristics and conditions of our individual and elemental ends and extend them *without correction* to life as a whole. Accident, disease and decrepitude spell the death of men; and therefore the same applies to mankind.

But have we any right to generalise in this simple way? When an individual disappears, even prematurely, another is always there to replace him. His loss is not irreparable from the point of view of the continuation of life. But what about mankind? In one of his books the great palaeontologist Matthew has suggested that if the human branch disappeared, another thinking branch would soon take its place. But he does not tell us where this mysterious shoot could be expected to appear on the tree of life as we know it, and doubtless he would be hard put to it to do so.

If we take the whole of history into consideration, the biological situation seems to me to be quite otherwise.

Once and once only in the course of its planetary existence has the earth been able to envelop itself with life. Similarly once and once only has life succeeded in crossing the threshold of reflection. For thought as for life there has been just one season. And we must not forget that since the birth of thought man has been the leading shoot of the tree of life. That being so, the hopes for the future of the noosphere (that is to say, of biogenesis, which in the end is the same as cosmogenesis) are concentrated exclusively upon him as such. How then could he come to an end before his time, or stop, or deteriorate, unless the universe committed abortion upon itself, which we have already decided to be absurd ?

In its present state, the world would be unintelligible and the presence in it of reflection would be incomprehensible, unless we supposed there to be a secret complicity between the infinite and the infinitesimal to warm, nourish and sustain to the very end—by dint of chance, contingencies and the exercise of free choice—the consciousness that has emerged between the two. It is upon this complicity that we must depend. *Man is irreplaceable.* Therefore, however improbable it might seem, *he must reach the goal*, not necessarily, doubtless, but infallibly.

What we should expect is not a halt in any shape or form, but an ultimate progress coming at its biologically appointed hour ; a maturation and a paroxysm leading ever higher into the Improbable from which we have sprung. It is in this direction that we must extrapolate man and hominisation if we want to get a forward glimpse of the end of the world.

## 2. THE APPROACHES

Without going beyond the limits of scientific probability, we can say that life still has before it long periods of geological time in which to develop. Moreover, in its thinking form, it still shows every sign of an energy in full expansion. On the one hand, compared with the zoological layers which preceded it whose average duration is at least in the order of eighty million years, mankind is so young that it could almost be called new-born. On the other hand, to judge from the rapid developments of thought in the short period of a few dozen centuries, this youth bears within it the indications and the promises of an entirely new biological cycle. Thus in all probability, between our modern earth and the ultimate earth, there stretches an immense period, characterised not by a slowing-down but a speeding up and by the definitive florescence of the forces of evolution along the line of the human shoot.

Assuming success—which is the only acceptable assumption—under what form and along what lines can we imagine progress developing during this period ?

In the first place, *in a collective and spiritual form.* We have noticed that, since man's advent, there has been a certain slowing down of the passive and somatic transformations of the organism in favour of the conscious and active metamorphoses of the individual absorbed in society. We find the artificial carrying on the work of the natural ; and the transmission of an oral or written culture being superimposed on genetic forms of heredity (chromosomes). Without denying the possibility or even probability of a certain prolongation in our limbs, and still more in our nervous system, of the orthogenetic processes of the past,[1] I am inclined to think that their influence, hardly appreciable since the emergence of *Homo sapiens*, is destined to dwindle still further. As thought regulated by a sort of quantum law, the energies of life seem unable to spread in one region or take on a new form except at the expense of a lowering elsewhere. Since man's arrival, the evolutionary pressure seems to have dropped in all the non-human branches of the tree of life. And now that man has become an adult and has opened up for himself the field of mental and social transformations, bodies no longer change appreciably ; they no longer need to in the

[1] Taken up again and prolonged reflectively, artificially—who knows ?—by biology (assault on the laws and springs of heredity, use of hormones, etc., see pp. 249–50).

human branch ; or if they still change, it will only be under our industrious control. It may well be that in its individual capacities and penetration our brain has reached its organic limits. But the movement does not stop there. From west to east, evolution is henceforth occupied elsewhere, in a richer and more complex domain, constructing, with all minds joined together, *mind.* Beyond all nations and races, the inevitable taking-as-a-whole of mankind has already begun.

With that said, we have now to ask : *along what lines* of advance, among others—judging from the present condition of the noosphere—are we destined to proceed from the planetary level of psychic totalisation and evolutionary upsurge we are now approaching ?

I can distinguish three principal ones in which we see again the predictions to which we were already led by our analysis of the ideas of science and humanity. They are : the organisation of research, the concentration of research upon the subject of man, and the conjunction of science and religion. These are three natural terms of one and the same progression.

### A. *The Organisation of Research*

We are given to boasting of our age being an age of science. And if we are thinking merely of the dawn compared to the darkness that went before, up to a point we are justified. Something enormous has been born in the universe with our discoveries and our methods of research. Something has been started which, I am convinced, will now never stop. Yet though we may exalt research and derive enormous benefit from it, with what pettiness of spirit, poverty of means and general haphazardness do we pursue truth in the world today ! Have we ever given serious thought to the predicament we are in ?

Like art—indeed we might almost say like thought itself—science was born with every sign of superfluity and fantasy. It was born of the exuberance of an internal activity that had outstripped the material needs of life ; it was born of the curiosity of dreamers and idlers. Gradually it became important ; its effectiveness gave it the freedom of the city. Living in a world which it can justly be said to have revolutionised, it has acquired a social status ; sometimes it is even worshipped. Yet we still leave it to grow as best it can, hardly tending it, like those wild plants whose fruits are plucked by primitive peoples in their forests. Everything is subordinated to the increase in industrial production, and to armaments. The scientist and the laboratories which multiply our powers still receive nothing, or next to nothing. We behave as though we expected discoveries to fall ready-made from the sky, like rain or sunshine, while men concentrate on the serious business of killing each other and eating. Let us stop to think for a moment of the proportion of human energy devoted, here and now, to the pursuit of truth. Or, in still more concrete terms, let us glance at the percentage of a nations' revenue allotted in its budget for the investigation of clearly-defined problems whose solution would be of vital consequence for the world. If we did we should be staggered. Less is provided annually for all the pure research all over the world than for one capital ship. Surely our great-grandsons will not be wrong if they think of us as barbarians ?

The truth is that, as children of a transition period, we are neither fully conscious of, nor in full control of, the new powers that have been unleashed. Clinging to outworn habit, we still see in science only a new means of providing more easily the same old things. We put Pegasus between the traces. And Pegasus languishes—unless he bolts with the waggon! But the moment will come—it is bound to—when man will be forced by disparity of the equipage to admit that science is not an accessory occupation for him but an essential activity, a natural derivative of the overspill of energy constantly liberated by mechanisation.

We can envisage a world whose constantly increasing 'leisure' and heightened interest would find their vital issue in fathoming everything, trying everything, extending everything ;

a world in which giant telescopes and atom smashers would absorb more money and excite more spontaneous admiration than all the bombs and cannons put together; a world in which, not only for the restricted band of paid research-workers, but also for the man in the street, the day's ideal would be the wresting of another secret or another force from corpuscles, stars, or organised matter ; a world in which, as happens already, one gives one's life to be and to know, rather than to possess. That, on an estimate of the forces engaged,[1] is what is being relentlessly prepared around us.

In some of the lower organisms the retina is, as it were, spread over the whole surface of the body. In somewhat the same way human vision is still diffuse in its operation, mixed up with industrial activity and war. Biologically it needs to individualise itself independently, with its own distinct organs. It will not be long now before the noosphere finds its eyes.

### B. *The Discovery of the Human Object*

When mankind has once realised that its first function is to penetrate, intellectually unify, and harness the energies which surround it, in order still further to understand and master them, there will no longer be any danger of running into an upper limit of its florescence. A commercial market can reach saturation point. One day, though substitutes may be found, we shall have exhausted our mines and oil-wells. But to all appearances nothing on earth will ever saturate our desire for knowledge or exhaust our power for invention. For of each may be said : *crescit eundo*.

That does not mean that science should propagate itself indifferently in any and every direction at the same time like a ripple in an isotropic medium. The more one looks, the more

[1] External forces of planetary compression obliging humanity to totalise itself organically in itself; and internal forces (ascendent and propulsive) of spiritualisation, unleashed or exalted by technico-social totalisation.

one sees. And the more one sees, the better one knows where to look. If life has been able to advance, it is because, by ceaseless groping, it has successively found the points of least resistance at which reality yielded to its thrust. Similarly, if research is to progress tomorrow, it will be largely by localising the central zones, the sensitive zones which are 'alive', whose conquest will afford us an easy mastery of all the rest.

From this point of view, if we are going towards a human era of science, it will be eminently an era of human science. Man, the knowing subject, will perceive at last that man, 'the object of knowledge', is the key to the whole science of nature.

Carrel referred to man as 'the unknown'. But man, we should add, is the solution of everything that we can know.

Up to the present, whether from prejudice or fear, science has been reluctant to look man in the face but has constantly circled round the human object without daring to tackle it. Materially our bodies seem insignificant, accidental, transitory and fragile ; why bother about them ? Psychologically, our souls are incredibly subtle and complex : how can one fit them into a world of laws and formulas?

Yet the more persistently we try to avoid man in our theories, the more tightly drawn become the circles we describe around him, as though we were caught up in his vortex. As I said in my Preface, at the end of its analyses, physics is no longer sure whether what is left in its hands is pure energy or, on the contrary, thought. At the end of its constructions, biology, if it takes its discoveries to their logical conclusion, finds itself forced to acknowledge the assemblage of thinking beings as the present terminal form of evolution. We find man at the bottom, man at the top, and, above all, man at the centre—man who lives and struggles desperately in us and around us. We shall have to come to grips with him sooner or later.

Man is, if I have not gone astray in these pages, an object of study of unique value to science for two reasons. (i) He represents, individually and socially, the most synthesised state under which the stuff of the universe is available to us. (ii)

Correlatively, he is at present the most mobile point of the stuff in course of transformation.

For these two reasons, to decipher man is essentially to try to find out how the world was made and how it ought to go on making itself. The science of man is the practical and theoretical science of hominisation. It means profound study of the past and of origins. But still more, it means constructive experiment pursued on a continually renewed object. The programme is immense and its only end or aim is that of the future.

What is involved, firstly, is the care and improvement of the human body, the health and strength of the organism. So long as its phase of immersion in the 'tangential' lasts, thought can only be built up on this material basis. And now, in the tumult of ideas that accompany the awakening of the mind, are we not undergoing physical degeneration ? It has been said that we might well blush comparing our own mankind, so full of misshapen subjects, with those animal societies in which, in a hundred thousand individuals, not one will be found lacking in a single antenna. In itself that geometrical perfection is not in the line of our evolution whose bent is towards suppleness and freedom. All the same, suitably subordinated to other values, it may well appear as an indication and a lesson. So far we have certainly allowed our race to develop at random, and we have given too little thought to the question of what medical and moral factors *must replace the crude forces of natural selection* should we suppress them. In the course of the coming centuries it is indispensable that a nobly human form of eugenics, on a standard worthy of our personalities, should be discovered and developed.

Eugenics applied to individuals leads to eugenics applied to society. It would be more convenient, and we would incline to think it safe, to leave the contours of that great body made of all our bodies to take shape on their own, influenced only by the automatic play of individual urges and whims. 'Better not interfere with the forces of the world !' Once more we are up against the mirage of instinct, the so-called infallibility of nature. But is it not precisely the world itself which, culminating in thought, expects us to think out again the instinctive impulses of nature so as to perfect them ? Reflective substance requires reflective treatment. If there is a future for mankind, it can only be imagined in terms of a harmonious conciliation of what is free with what is planned and totalised. Points involved are : the distribution of the resources of the globe ; the control of the trek towards unpopulated areas ; the optimum use of the powers set free by mechanisation ; the physiology of nations and races; geo-economy, geo-politics, geo-demography; the organisation of research developing into a reasoned organisation of the earth. Whether we like it or not, all the signs and all our needs converge in the same direction. We need and are irresistibly being led to create, by means of and beyond all physics, all biology and all psychology, *a science of human energetics.*

It is in the course of that creation, already obscurely begun, that science, by being led to concentrate on man, will find itself increasingly face to face with religion.

### C. *The Conjunction of Science and Religion*

To outward appearance, the modern world was born of an anti-religious movement: man becoming self-sufficient and reason supplanting belief. Our generation and the two that preceded it have heard little but talk of the conflict between science and faith ; indeed it seemed at one moment a foregone conclusion that the former was destined to take the place of the latter.

But, as the tension is prolonged, the conflict visibly seems to need to be resolved in terms of an entirely different form of equilibrium—not in elimination, nor duality, but in synthesis. After close on two centuries of passionate struggles, neither science nor faith has succeeded in discrediting its adversary. On the contrary, it becomes obvious that neither can develop normally without the other. And the reason is simple : the same life animates both. Neither in its impetus nor its achievements can science go to its limits without becoming tinged with mysticism and charged with faith,

Firstly *in its impetus.* We touched on this point when dealing with the problem of action. Man will only continue to work and to research so long as he is prompted by a passionate interest. Now this interest is entirely dependent on the conviction, strictly undemonstrable to science, that the universe has a direction and that it could—indeed, if we are faithful, it *should*—result in some sort of irreversible perfection. Hence comes belief in progress.

Secondly *in its construction.* Scientifically we can envisage an almost indefinite improvement in the human organism and human society. But as soon as we try to put our dreams into practice, we realise that the problem remains indeterminate or even insoluble unless, with some partially super-rational intuition, we admit the convergent properties of the world we belong to. Hence belief in unity.

Furthermore, if we decide, under the pressure of facts, in favour of an optimism of unification, we run into the technical necessity of discovering—in addition to the impetus required to push us forward and in addition to the particular objective which should determine our route—the special binder or cement which will associate our lives together, vitally, without diminishing or distorting them. Hence, belief in a supremely attractive centre which has personality.

In short, as soon as science outgrows the analytic investigations which constitute its lower and preliminary stages, and passes on to synthesis—synthesis which naturally culminates in the realisation of some superior state of humanity—it is at once led to foresee and place its stakes on the *future* and on the *all.* And with that it out-distances itself and emerges in terms of *option* and *adoration.*

Thus Renan and the nineteenth century were not wrong to speak of a Religion of Science. Their mistake was not to see that their cult of humanity implied the re-integration, in a renewed form, of those very spiritual forces they claimed to be getting rid of.

When, in the universe in movement to which we have just awakened, we look at the temporal and spatial series diverging and amplifying themselves around and behind us like the laminae of a cone, we are perhaps engaging in pure science. But when we turn towards the summit, towards the *totality* and the *future*, we cannot help engaging in religion.

Religion and science are the two conjugated faces or phases of one and the same complete act of knowledge—the only one which can embrace the past and future of evolution so as to contemplate, measure and fulfil them.

In the mutual reinforcement of these two still opposed powers, in the conjunction of reason and mysticism, the human spirit is destined, by the very nature of its development, to find the uttermost degree of its penetration with the maximum of its vital force.

## 3. THE ULTIMATE

Always pushing forward in the three directions we have just indicated, and taking advantage of the immense duration it has still to live, mankind has enormous possibilities before it.

Until the coming of man, life was quickly arrested and hemmed in by the specialisations into which it was forced to mould itself so as to act, and became fixed, then dispersed, at each forward bound. Since the threshold of reflection, we have entered into an entirely new field of evolution—thanks to the astonishing properties of 'artifice' which separate the instrument from the organ and enable one and the same creature to intensify and vary the modalities of its action indefinitely without losing anything of its freedom ; and thanks to the prodigious power of thought to bring together and combine in a single conscious effort all the human particles. In fact, though the study of the past may give us some idea of the resources of organised matter in its dispersed state, *we have as yet no idea of the possible magnitude* of 'noospheric' effects. We are confronted with human vibrations resounding by the million—a whole layer of consciousness exerting simultaneous pressure upon the future and the collected and hoarded produce of a million years of thought. Have we ever tried to form an idea of what such magnitudes represent ?[1]

In this direction, the most unexpected is perhaps what we should most expect. Under the increasing tension of the mind on the surface of the globe, we may begin by asking seriously whether life will not perhaps one day succeed in ingeniously forcing the bars of its earthly prison, either by finding the means to invade other inhabited planets or (a still more giddy perspective) by getting into psychical touch with other focal points of consciousness across the abysses of space. The meeting and mutual fecundation of two noospheres is a supposition which may seem at first sight crazy, but which after all is merely extending to psychical phenomena a scope no-one would think of denying to material phenomena. Consciousness would thus finally construct itself by a synthesis of planetary units. Why not, in a universe whose astral unit is the galaxy ?

Without in any way wishing to discourage such hypotheses —whose realisation, though enormously enlarging the dimensions, would leave unchanged both the convergent form and hence the final duration of noogenesis—I consider their probability too remote for them to be worth dwelling on.

The human organism is so extraordinarily complicated and

[1] Over and above the intellectual value of isolated human units, there are thus grounds for recognising a collective exaltation (by mutual support or reverberation) when those units are suitably arranged. It would be difficult to say whether there are any Aristotles, Platos or St. Augustines now on earth (how could it be proved : on the other hand why not?) But what is clear is that, each supporting the other (making a single arch or a single mirror), our modern souls see and feel today a world such as (in size, inter-connections and potentialities) escaped all the great men of antiquity. To this progress in consciousness, could anyone dare to object that there has been no corresponding advance in the profound structure of being?

sensitive, and so closely adapted to terrestrial conditions, that it is difficult to see how man could acclimatise himself to another planet, even if he were capable of navigating through interplanetary space. The sidereal durations are so immense that it is difficult to see how in two different regions of the heavens, two thought systems could co-exist and coincide at comparable stages of their development. For these two reasons among others I adopt the supposition that our noosphere is destined to close in upon itself in isolation, and that it is in a psychical rather than a spatial direction that it will find an outlet, without need to leave or overflow the earth. Hence, quite naturally, the notion of change of state recurs.

Noogenesis rises upwards in us and through us unceasingly. We have pointed to the principal characteristics of that movement : the closer association of the grains of thought ; the synthesis of individuals and of nations or races ; the need of an autonomous and supreme personal focus to bind elementary personalities together, without deforming them, in an atmosphere of active sympathy. And, once again : all this results from the combined action of two curvatures—the roundness of the earth and the cosmic convergence of mind—in conformity with the law of complexity and consciousness.

Now when sufficient elements have sufficiently agglomerated, this essentially convergent movement will attain such intensity and such quality that mankind, *taken as a whole*, will be obliged—as happened to the individual forces of instinct—to reflect upon itself at a single point ;[1] that is to say, in this case, to abandon its organo-planetary foothold so as to shift its centre on to the transcendent centre of its increasing concentration. This will be the end and the fulfilment of the spirit of the earth.

The end of the world : the wholesale internal introversion upon itself of the noosphere, which has simultaneously reached the uttermost limit of its complexity and its centrality.

The end of the world : the overthrow of equilibrium,

[1] Which amounts to saying that human history develops between two points of reflection, the one inferior and individual, the other superior and collective.

detaching the mind, fulfilled at last, from its material matrix, so that it will henceforth rest with all its weight on God-Omega.

The end of the world : critical point simultaneously of emergence and emersion, of maturation and escape.

We can entertain two almost contradictory suppositions about the physical and psychical state our planet will be in as it approaches maturation.[1] According to the first hypothesis which expresses the hopes towards which we ought in any case to turn our efforts as to an ideal, evil on the earth at its final stage will be reduced to a minimum. Disease and hunger will be conquered by science and we will no longer need to fear them in any acute form. And, conquered by the sense of the earth and human sense, hatred and internecine struggles will have disappeared in the ever-warmer radiance of Omega. Some sort of unanimity will reign over the entire mass of the noosphere. The final convergence will take place *in peace*.[2] Such an outcome would of course conform most harmoniously with our theory.

But there is another possibility. Obeying a law from which nothing in the past has ever been exempt, evil may go on growing alongside good, and it too may attain its paroxysm at the end in some specifically new form.

There are no summits without abysses.

Enormous powers will be liberated in mankind by the inner play of its cohesion : though it may be that this energy will still be employed discordantly tomorrow, as today and in the past. Are we to foresee a mechanising synergy under brute force, or a synergy of sympathy ? Are we to foresee man seeking to fulfil himself collectively upon himself, or personally on a greater than himself ? Refusal or acceptance of Omega? A conflict may supervene. In that case the noosphere, in the course of and by virtue of the process which draws it together, will,

[1] On the degree of 'inevitability' of this maturation of a *free* mass, see Conclusion, p. 309.

[2] Though at the same time—since a critical point is being approached—*in extreme tension.* There is nothing in common between this perspective and the old millenary dreams of a terrestrial paradise at the end of time.

when it has reached its point of unification, split into two zones each attracted to an opposite pole of adoration. Thought has never completely united upon itself here below. Universal love would only vivify and detach finally a fraction of the noosphere so as to consummate it—the part which decided to 'cross the threshold', to get outside itself into the other. *Ramification once again, for the last time.*

In this second hypothesis, which is more in conformity with traditional apocalyptic thinking, we may perhaps discern three curves around us rising up at one and the same time into the future : an inevitable education in the organic possibilities of the earth, an internal schism of consciousness ever increasingly divided on two opposite ideals of evolution, and positive attraction of the centre of centres at the heart of those who turn towards it. And the earth would finish at the triple point at which, by a coincidence altogether in keeping with the ways of life, these three curves would meet and attain their maximum at the very same moment.

The death of the materially exhausted planet ; the split of the noosphere, divided on the form to be given to its unity ; and simultaneously (endowing the event with all its significance and with all its value) the liberation of that percentage of the universe which, across time, space and evil, will have succeeded in laboriously synthesising itself to the very end.

Not an indefinite progress, which is an hypothesis contradicted by the convergent nature of noogenesis, but an ecstasy transcending the dimensions and the framework of the visible universe.

Ecstasy in concord ; or discord ; but in either case by excess of interior tension : the only biological outcome proper to or conceivable for the phenomenon of man.

Among those who have attempted to read this book to the end, many will close it, dissatisfied and thoughtful, wondering whether I have been leading them through facts, through metaphysics or through dreams.

But have those who still hesitate in this way really understood the rigorous and salutary conditions imposed on our reason by the coherence of the universe, now admitted by all ? A mark appearing on a film; an electroscope discharging abnormally ; that is enough to force physics to accept fantastic powers in the atom. Similarly, if we try to bring man, body and soul, within the framework of what is experimental, man obliges us to readjust completely to his measure the layers of time and space.

To make room for thought in the world, I have needed to 'interiorise' matter : to imagine an energetics of the mind ; to conceive a noogenesis rising upstream against the flow of entropy ; to provide evolution with a direction, a line of advance and critical points ; and finally to make all things double back upon *someone*.

In this arrangement of values I may have gone astray at many points. It is up to others to try to do better. My one hope is that I have made the reader feel both the reality, difficulty, and urgency of the problem and, at the same time, the scale and the form which the solution cannot escape.

The only universe capable of containing the human person is an irreversibly 'personalising' universe.

EPILOGUE

# THE CHRISTIAN PHENOMENON

NEITHER IN the play of its elemental activities, which can only be set in motion by the hope of an 'imperishable'; nor in the play of its collective affinities, which require for their coalescence the action of a conquering love, can reflective life continue to function and to progress unless, above it, there is a pole which is supreme in attraction and consistence. By its very structure the noosphere could not close itself either individually or socially in any way save under the influence of the centre we have called Omega.

That is the postulate to which we have been led logically by the integral application to man of the experimental laws of evolution. The possible, or even the probable, repercussion of this conclusion, however theoretical in the first approximation, upon experience will now be obvious.

If Omega were only a remote and ideal focus destined to emerge at the end of time from the convergence of terrestrial consciousnesses, nothing could make it known to us but this convergence. At the present time no other energy of a personal nature could be detected on earth save that represented by the sum of human persons.

If, on the other hand, Omega is, as we have admitted, *already in existence* and operative at the very core of the thinking mass, then it would seem inevitable that its existence should be manifested to us here and now through some traces. To animate evolution in its lower stages, the conscious pole of the world could of course only act in an impersonal form and under the veil of biology. Upon the thinking entity that we have become by hominisation, it is now possible for it to radiate from the one centre to all centres—*personally*. Would it seem likely that it should not do so ?

Either the whole construction of the world presented here is vain ideology or, somewhere around us, in one form or another, some excess of personal, extra-human energy should be perceptible to us if we look carefully, and should reveal to us the great Presence. It is at this point that we see the importance for science of *the Christian phenomenon.*

At the conclusion of a study of *the human phenomenon* I have not chosen those words haphazardly, nor for the sake of mere verbal symmetry. They are meant to define without ambiguity the spirit in which I want to speak.

As I am living at the heart of the Christian world, I might be suspected of wanting to introduce an apologia by artifice. But, here again, so far as it is possible for a man to separate in himself the various planes of knowledge, it is not the convinced believer but the naturalist who is asking for a hearing.

The Christian fact stands before us. It has its place among the other realities of the world.

I would like to show how it seems to me to bring to the perspectives of a universe dominated by energies of a personal nature the crucial confirmation we are in need of, firstly by the substance of its creed, next, by its existence-value, and finally by its extraordinary power of growth.

## 1. AXES OF BELIEF

To those who only know it outwardly, Christianity seems desperately intricate. In reality, taken in its main lines, it contains an extremely simple and astonishingly bold solution of the world.

In the centre, so glaring as to be disconcerting, is the uncompromising affirmation of a personal God : God as providence, directing the universe with loving, watchful care ; and God the revealer, communicating himself to man on the level of and through the ways of intelligence. It will be easy for me, after all I have said, to demonstrate the value and actuality of this tenacious personalism, not long since condemned as obsolete. The important thing to point out here is the way in which such an attitude in the hearts of the faithful leaves the door open to, and is easily allied to, everything that is great and healthy in the universal.

In its Judaic phase, Christianity might well have considered itself the particular religion of one people. Later on, coming under the general conditions of human knowledge, it came to think that the world around it was much too small. However that may be, it was hardly constituted before it was ceaselessly trying to englobe in its constructions and conquests the totality of the system that it managed to picture to itself.

Personalism and universalism : in what form have these two characters been able to unite in its theology ?

For reasons of practical convenience and perhaps also of intellectual timidity, the City of God is too often described in pious works in conventional and purely moral terms. God and the world he governs are seen as a vast association, essentially legalistic in its nature, conceived in terms of a family or government. The fundamental root from which the sap of Christianity has risen from the beginning and is nourished, is quite otherwise. Led astray by a false evangelism, people often think they are honouring Christianity when they reduce it to a sort of gentle philanthropism. Those who fail to see in it the most realistic and at the same time the most cosmic of beliefs and hopes, completely fail to understand its ' mysteries '. Is the Kingdom of God a big family ? Yes, in a sense it is. But in another sense it is a prodigious biological operation—that of the Redeeming Incarnation.

As early as in St. Paul and St. John we read that to create, to fulfil and to purify the world is, for God, to unify it by uniting it organically with himself.[1] How does he unify it ? By partially immersing himself in things, by becoming 'element', and then, from this point of vantage in the heart of matter, assuming the control and leadership of what we now call evolution. Christ, principle of universal vitality because sprung up as man among men, put himself in the position (maintained ever since) to subdue under himself, to purify, to direct and superanimate the general ascent of consciousnesses into which he inserted himself. By a perennial act of communion and sublimation, he aggregates to himself the total psychism of the earth. And when he has gathered everything together and transformed everything, he will close in upon himself and his conquests, thereby rejoining, in a final gesture, the divine focus he has never left. Then, as St. Paul tells us, *God shall be all in all.* This is indeed a superior form of 'pantheism'[2] without trace of the poison of adulteration or annihilation : the expectation of perfect unity, steeped in which each element will reach its consummation at the same time as the universe.

The universe fulfilling itself in a synthesis of centres in perfect conformity with the laws of union. God, the Centre of centres. In that final vision the Christian dogma culminates. And so exactly, so perfectly does this coincide with the Omega Point that doubtless I should never have ventured to envisage the latter or formulate the hypothesis rationally if, in my consciousness as a believer, I had not found not only its speculative model but also its living reality.

## 2. EXISTENCE VALUE

It is relatively easy to build up a theory of the world. But it is beyond the powers of an individual to provoke artificially the

[1] Following Greek thought—following all thought in fact—are not 'to be' and 'to be one' identical ?

[2] '*En pâsi panta Theos.*'

birth of a religion. Plato, Spinoza and Hegel were able to elaborate views which compete in amplitude with the perspectives of the Incarnation. Yet none of these metaphysical systems advanced beyond the limits of an ideology. Each in turn has perhaps brought light to men's minds, but without ever succeeding in begetting life. What to the eyes of a 'naturalist' comprises the importance and the enigma of the Christian phenomenon is its existence-value and reality-value.

Christianity is in the first place real by virtue of the spontaneous amplitude of the movement it has managed to create in mankind. It addresses itself to every man and to every class of man, and from the start it took its place as one of the most vigorous and fruitful currents the noosphere has ever known. Whether we adhere to it or break off from it, we are surely obliged to admit that its stamp and its enduring influence are apparent in every corner of the earth today.

It is doubtless a quantitative value of life if measured by its radius of action; but it is still more a qualitative value which expresses itself—like all biological progress—by the appearance of a specifically new state of consciousness.

I am thinking here of Christian love.

Christian love is incomprehensible to those who have not experienced it. That the infinite and the intangible can be lovable, or that the human heart can beat with genuine charity for a fellow-being, seems impossible to many people I know —in fact almost monstrous. But whether it be founded on an illusion or not, how can we doubt that such a sentiment exists, and even in great intensity ? We have only to note crudely the results it produces unceasingly all round us. Is it not a positive fact that thousands of mystics, for twenty centuries, have drawn from its flame a passionate fervour that outstrips by far in brightness and purity the urge and devotion of any human love ? is it not also a fact that, having once experienced it, further thousands of men and women are daily renouncing every other ambition and every other joy save that of abandoning themselves to it and labouring within it more and more completely ?

Lastly, is it not a fact, as I can warrant, that if the love of God were extinguished in the souls of the faithful, the enormous edifice of rites, of hierarchy and of doctrines that comprise the Church would instantly revert to the dust from which it rose ?

It is a phenomenon of capital importance for the science of man that, over an appreciable region of the earth, a zone of thought has appeared and grown in which a genuine universal love has not only been conceived and preached, but has also been shown to be psychologically possible and operative in practice. It is all the more capital inasmuch as, far from decreasing, the movement seems to wish to gain still greater speed and intensity.

## 3. POWER OF GROWTH

For almost all the ancient religions, the renewal of cosmic outlook characterising 'the modern mind' has occasioned a crisis of such severity that, if they have not yet been killed by it, it is plain they will never recover. Narrowly bound to untenable myths, or steeped in a pessimistic and passive mysticism, they can adjust themselves neither to the precise immensities, nor to the constructive requirements, of space-time. They are out of step both with our science and with our activity.

But under the shock which is rapidly causing its rivals to disappear, Christianity, which might at first have been thought to be shaken too, is showing, on the contrary, every sign of forging ahead. For, by the very fact of the new dimensions assumed by the universe as we see it today, it reveals itself both as inherently more vigorous in itself and as more necessary to the world than it has ever been before.

*More vigorous.* To live and develop the Christian outlook needs an atmosphere of greatness and of coherence. The bigger the world becomes and the more organic become its internal connections, the more will the perspectives of the Incarnation triumph. That is what believers are beginning, much to their surprise, to find out. Though frightened for a moment by evolution, the Christian now perceives that what it offers him is nothing but a magnificent means of feeling more at one with God and of giving himself more to him. In a pluralistic and static Nature, the universal domination of Christ could, strictly speaking, still be regarded as an extrinsic and super-imposed power. In a spiritually converging world this 'Christic' energy acquires an urgency and intensity of another order altogether. If the world is convergent and if Christ occupies its centre, then the Christogenesis of St. Paul and St. John is nothing else and nothing less than the extension, both awaited and unhoped for, of that noogenesis in which cosmogenesis—as regards our experience—culminates. Christ invests himself organically with the very majesty of his creation. And it is in no way metaphorical to say that man finds himself capable of experiencing and discovering his God in the whole length, breadth and depth of the world in movement. To be able to say literally to God that one loves him, not only with all one's body, all one's heart and all one's soul, but with every fibre of the unifying universe—that is a prayer that can only be made in space-time.

*More necessary*. To say of Christianity that, despite appearances to the contrary, it is acclimatising itself and expanding in a world enormously enlarged by science, is to point to no more than one half of the picture. Evolution has come to infuse new blood, so to speak, into the perspectives and aspirations of Christianity. In return, is not the Christian faith destined, is it not preparing, to save and even to take the place of evolution ?

I have tried to show that we can hope for no progress on earth without the primacy and triumph of the *personal* at the summit of *mind*. And at the present moment Christianity is the *unique* current of thought, on the entire surface of the noosphere, which is sufficiently audacious and sufficiently progressive to lay hold of the world, at the level of effectual practice, in an embrace, at once already complete, yet capable of indefinite perfection, where faith and hope reach their fulfilment in love. *Alone*, unconditionally alone, in the world today, Christianity shows itself able to reconcile, in a single living act, the All and the Person. Alone, it can bend our hearts not only to the service of that tremendous movement of the world which bears us along, but beyond, to embrace that movement in love.

In other words can we not say that Christianity fulfils all the conditions we are entitled to expect from a religion of the future ; and that hence, through it, the principal axis of evolution truly passes, as it maintains ?

Now let us sum up the situation :

i. Considered objectively asa phenomenon, the Christian movement, through its rootedness in the past and ceaseless developments, exhibits the characteristics of a *phylum*.

ii. Reset in an evolution interpreted as an ascent of consciousness, this phylum, in its trend towards a synthesis based on love, progresses precisely in the direction presumed for the leading-shoot of biogenesis.

iii. In the impetus which guides and sustains its advance, this rising shoot implies essentially *the consciousness of being in actual relationship* with a spiritual and transcendent pole of universal convergence.

To confirm the presence at the summit of the world of what we have called the Omega Point,[1] do we not find here the very cross-check we were waiting for ? Here surely is the ray of sunshine striking through the clouds, the reflection onto what is ascending of that which is already on high, the rupture of our solitude. The palpable influence on our world of *an other* and supreme Someone . . . Is not the Christian phenomenon, which

[1] To be more exact, ' to confirm the presence at the summit of the world of something in line with, but still more elevated than, the Omega point '. This is in deference to the theological concept of the ' supernatural ' according to which the binding contact between God and the world, *hic et nunc* inchoate, attains to a super-intimacy (hence also a super-gratuitousness) of which man can have no inkling and to which he can lay no claim by virtue of his ' nature ' alone.

rises upwards at the heart of the social phenomenon, precisely that?

In the presence of such perfection in coincidence, even if I were not a Christian but only a man of science, I think I would ask myself this question.

*Peking, June 1938–June 1940*

SUMMING UP OR POSTSCRIPT

# THE ESSENCE OF THE PHENOMENON OF MAN

SINCE THIS book was composed, I have experienced no change in the intuition it seeks to express. Taken as a whole, I still see man today exactly as I saw him when I first wrote these pages. Yet the basic vision has not remained—it could not remain—stationary. By the irresistible deepening of reflection, by the decantation and automatic patterning of associated ideas, by the discovery of new facts and by the continual need to be better understood, certain new formulations and articulations have gradually occurred to me in the last ten years. They tend to emphasise, and at the same time to simplify, the main lines of my earlier draft.

It is this unchanged, though recogitated, essence of the *Phenomenon of Man* which I think it will be useful to set out here as a summing-up or conclusion under three inter-related headings:

## 1. A WORLD IN INVOLUTION, OR THE COSMIC LAW OF COMPLEXITY-CONSCIOUSNESS

The astronomers have lately been making us familiar with the idea of a universe which for the last few thousand million years has been expanding in galaxies from a sort of primordial atom. This perspective of a world in a state of explosion is still debated, but no physicist would think of rejecting it as being tainted with philosophy or finalism. The reader should keep this example before him when he comes to weigh up the scope, the limitations and the perfect scientific legitimacy of the views I have here put forward. Reduced to its ultimate essence, the substance of these long pages can be summed up in this simple affirmation : that if the universe, regarded sidereally, is in process of spatial expansion (from the infinitesimal to the immense), in the same way and still more clearly it presents itself to us, physico-chemically, as in process of organic *involution* upon itself (from the extremely simple to the extremely complex)—and, moreover, this particular involution 'of complexity' is experimentally bound up with a correlative increase in interiorisation, that is to say in the psyche or consciousness.

In the narrow domain of our planet (still the only one within the scope of biology) the structural relationship noted here between complexity and consciousness is experimentally incontestable and has always been known. What gives the standpoint taken in this book its originality is the affirmation, at the outset, that the particular property possessed by terrestrial substances—of becoming more vitalised as they become increasingly complex—is only the local manifestation and expression of a trend as universal as (and no doubt even more significant than) those already identified by science : those trends which cause the cosmic layers not only to expand explosively as a wave but also to condense into corpuscles under the action of electromagnetic and gravitational forces, or perhaps to become dematerialised in radiation : trends which are probably strictly inter-connected, as we shall one day realise.

If that be so, it will be seen that consciousness (defined experimentally as the specific effect of organised complexity) transcends by far the ridiculously narrow limits within which our eyes can directly perceive it.

On the one hand we are logically forced to assume the existence in rudimentary form (in a microscopic, i.e. an infinitely diffuse, state) of some sort of psyche in every corpuscle, even in those (the mega-molecules and below) whose complexity is of such a low or modest order as to render it (the psyche) imperceptible—just as the physicist assumes and can calculate those changes of mass (utterly imperceptible to direct observation) occasioned by slow movement.

On the other hand, there precisely in the world where various physical conditions (temperature, gravity, etc.) prevent complexity reaching a degree involving a perceptible radiation of consciousness, we are led to assume that the involution, temporarily halted, will resume its advance as soon as conditions are favourable.

Regarded along its axis of complexity, the universe is, both on the whole and at each of its points, in a continual tension of organic doubling-back upon itself, and thus of interiorisation. Which amounts to saying that, for science, life is always under pressure everywhere ; and that where it has succeeded in breaking through in an appreciable degree, nothing will be able to stop it carrying to the uttermost limit the process from which it has sprung.

It is in my opinion necessary to take one's stand in this actively convergent cosmic setting if one wants to depict the phenomenon of man in its proper relief and explain it fully and coherently.

## 2. THE FIRST APPEARANCE OF MAN, OR THE INDIVIDUAL THRESHOLD OF REFLECTION

So as to overcome the improbability of arrangements leading to units of ever increasing complexity, the involuting universe, considered in its pre-reflective zones,[1] proceeds step by step by dint of billion-fold trial and error. It is this process of groping, combined with the two-fold mechanism of reproduction and heredity (allowing the hoarding and the additive improvement

[1] Once the threshold of reflection is crossed, the play of 'planned' or 'invented' combinations come into the picture, and to some extent supplants that of fortuitous combinations that 'just happen'. See below.

of favourable combinations obtained, without the diminution, indeed with the increase, of the number of individuals engaged), which gives rise to the extraordinary assemblage of living stems forming what I have called the tree of life—though I could equally well have chosen another image, that of the spectrum, in which each wavelength would correspond to a particular shade of consciousness or instinct.

From one point of view, the various stems of this psychical fan may seem (indeed they are often so regarded by science) to be vitally equivalent—just so many instincts, so many equally valid solutions to a given problem, comparison between which is futile. A second original point in my position in *The Phenomenon of Man*—apart from the interpretation of life as a universal function of the cosmos—lies, on the contrary, in giving the appearance on the human line of the power of reflection the value of a 'threshold' or a change of state. This affirmation is far from being an unwarranted assumption or based initially on any metaphysics of thought. It is a choice depending experimentally on the curiously underestimated fact that, from the threshold of reflection onwards, we are at what is nothing less than a new form of biological existence,[1] characterised, amongst other peculiarities, by the following properties:

*a.* The decisive emergence in individual life of factors of internal arrangement (*invention*) above the factors of external arrangement (utilisation of the play of chance).

*b.* The equally decisive appearance between elements of true forces of attraction and repulsion (sympathy and antipathy), replacing the pseudo-attractions and pseudo-repulsions of pre-life or even of the lower forms of life, which we seem to be able to refer back to simple reactions to the curves of space-time in the one case, and of the biosphere in the other.

[1] In exactly the same way as physics changes (with the introduction and dominance of certain new terms) when it passes from the scale of the medium-sized to that of the immense or, on the other hand, to that of the infinitesimal. It is too often forgotten that there should be, and is, a special biology of the 'infinitely complex'.

*c.* Lastly, the awakening in the consciousness of each particular element (consequent upon its new and revolutionary aptitude for foreseeing the future) of a demand for 'unlimited survival'. That is to say, the passage, for life, from a state of relative irreversibility (the physical impossibility of the cosmic involution to stop, once it has begun) to a state of absolute irreversibility (the radical dynamic incompatibility of a certain prospect of total death with the continuation of an evolution that has become reflective).

These various properties confer on the zoological group possessing them a superiority that is not only quantitative and numerical, but functional and vital—an indisputable superiority, I maintain, provided that we make up our minds to apply relentlessly and to the bitter end the experimental law of Complexity-Consciousness to the global evolution of the entire group.

## 3. THE SOCIAL PHENOMENON OR THE ASCENT TOWARDS A COLLECTIVE THRESHOLD OF REFLECTION

As we have seen, from a purely descriptive point of view, man was originally only one of innumerable branches forming the anatomic and psychic ramifications of life. But because this particular stem, or radius, alone among others, has succeeded, thanks to a privileged structure or position, in emerging from instinct into thought, it proves itself capable of spreading out in its turn, within this still completely free zone of the world, so as to form a spectrum of another order—the immense variety of anthropological types known to us. Let us take a glance at this second fanning-out. In virtue of the particular form of cosmogenesis adopted here, the problem our existence sets before our science is plainly the following: To what extent and eventually under what form does the human layer still obey (or is exempt from) the forces of cosmic involution which gave it birth?

The answer to this question is vital for our conduct, and depends entirely on the idea we form (or rather ought to form) of the nature of the social phenomenon as we now see it in full impetus around us.

As a matter of intellectual routine and because of the positive difficulty of mastering a process in which we are ourselves swept along, the constantly increasing auto-organisation of the human myriad upon itself is still regarded more often than not as a juridical or accidental process only superficially, 'extrinsically', comparable with those of biology. Naturally, it is admitted, mankind has always been increasing, which forces it to make more and more complex arrangements for its members. But these *modus vivendi* must not be confused with genuine ontological progress. From an evolutionary point of view, man has stopped moving, if he ever did move.

And this is where, as a man of science, I feel obliged to make my protest and object.

A certain sort of common sense[1] tells us that with man biological evolution has reached its ceiling : in reflecting upon itself, life has become stationary. But should we not rather say that it leaps forward ? Look at the way in which, as mankind technically patterns its multitudes, *pari passu* the psychic tension within it increases, with the consciousness of time and space and the taste for, and power of, discovery. This great event we accept without surprise. Yet how can one fail to recognise this revealing association of technical organisation and inward spiritual concentration as the work of the same great force (though in proportions and with a depth hitherto never attained), the very force which brought us into being ? How can we fail to see that after rolling us on individually—all of us, you and me—upon our own selves, it is still the same cyclone (only now on the social scale) which is still blowing over our heads, driving us together into a contact which tends to perfect each one of us by linking him organically to each and all of his neighbours ?

[1] The same 'common sense' which has again and again been corrected beyond all question by physics.

'Through human socialisation, whose specific effect is to involute upon itself the whole bundle of reflexive scales and fibres of the earth, it is the very axis of the cosmic vortex of interiorisation which is pursuing its course': replacing and extending the two preliminary postulates stated above (the one concerning the primacy of life in the universe, the other the primacy of reflection in life) this is the third option—the most decisive of all—which completes the definition and clarification of my scientific position as regards the phenomenon of man.

This is not the place to show in detail how easily and coherently this organic interpretation of the social phenomenon explains, or even in some directions allows us to predict, the course of history. Let it merely be stated that, if above the elementary hominisation that culminates in each individual, there is really developing above us another hominisation, a collective one of the whole species, then it is quite natural to observe, parallel with the socialisation of humanity, the same three psycho-biological properties rising upwards on the earth that the individual step to reflection originally produced.

*a.* Firstly the power of invention, so rapidly intensified at the present time by the rationalised collaboration of all the forces of research that it is already possible to speak of a human rebound of evolution.

*b.* Next, capacity for attraction (or repulsion), still operating in a chaotic way throughout the world but rising so rapidly around us that (whatever be said to the contrary) economics will soon count for very little in comparison with the ideological and the emotional factors in the arrangement of the world.

*c.* Lastly and above all, the demand for irreversibility. This emerges from the still somewhat hesitating zone of individual aspirations, so as to find categorical expression in consciousness and through the voice of the species. Categorical in the sense that, if an isolated man can succeed in imagining that it is possible physically, or even morally, for him to contemplate a complete suppression of himself—confronted with a total annihilation (or even simply with an insufficient preservation) destined for the fruit of his evolutionary labour—mankind, in its turn, is beginning to realise once and for all that its only course would be to go on strike. For the effort to push the earth forward is much too heavy, and the task threatens to go on much too long, for us to continue to accept it, unless we are to work in what is incorruptible.

These and other assembled pointers seem to me to constitute a serious scientific proof that (in conformity with the universal law of centro-complexity) the zoological group of mankind—far from drifting biologically, under the influence of exaggerated individualism, towards a state of growing granulation ; far from turning (through space-travel) to an escape from death by sidereal expansion ; or yet again far from simply declining towards a catastrophe or senility—the human group is in fact turning, by planetary arrangement and convergence of all elemental terrestrial reflections, towards a second critical pole of reflection of a collective and higher order ; towards a point beyond which (precisely because it is critical) we can see nothing directly, but a point through which we can nevertheless prognosticate the contact between thought, born of involution upon itself of the stuff of the universe, and that transcendent focus we call Omega, the principle which at one and the same time makes this involution irreversible and moves and gathers it in.

It only remains for me, in bringing this work to a close, to define my opinion on three matters which usually puzzle my readers : (a) what place remains for freedom (and hence for the possibility of a setback in the world) ? (b) what value must be given to spirit (as opposed to matter) ? and (c) what is the distinction between God and the World in the theory of cosmic involution ?

*a.* As regards the chances of success of cosmogenesis, my contention is that it in no way follows from the position taken up here that the final success of hominisation is necessary, inevitable and certain. Without doubt, the 'noogenic' forces of compression, organisation and interiorisation, under which the biological synthesis of reflection operates, do not at any moment relax their pressure on the stuff of mankind. Hence the possibility of foreseeing with certainty (*if all goes well*) certain precise directions of the future.[1] But, in virtue of its very nature, as we must not forget, the arrangement of great complexes (that is to say, of states of greater and greater improbability, even though closely linked together) does not operate in the universe (least of all in man) except by two related methods : (i) the groping utilisation of favourable cases (whose appearance is provoked by the play of large numbers) and (ii) in a second phase, reflective invention. And what does that amount to if not that, however persistent and imperious the cosmic energy of involution may be in its activity, it finds itself intrinsically influenced in its effects by two uncertainties related to the double play—chance at the bottom and freedom at the top ? Let me add, however, that in the case of very large numbers (such, for instance, as the human population) the process tends to 'unfallibilise' itself, inasmuch as the likelihood of success grows on the lower side (chance) while that of rejection and error diminishes on the other side (freedom) with the multiplication of the elements engaged.[2]

*b*. As regards the value of the spirit, I would like to say that from the phenomenal point of view, to which I systematically confine myself, matter and spirit do not present themselves as 'things' or 'natures' but as simple related *variables*, of which it behoves us to determine not the secret essence but the curve in function of space and time. And I recall that at this level of reflection 'consciousness' presents itself and demands to be treated, not as a sort of particular and subsistent entity, but as an 'effect', as the 'specific effect' of complexity.

[1] This for instance; that nothing could stop man in his advance to social unification, towards the development of machinery and automation (liberators of the spirit), towards 'trying all' and 'thinking all' right to the very end.

[2] For a Christian believer it is interesting to note that the final success of hominisation (and thus cosmic involution) is positively guaranteed by the 'redeeming virtue' of the God incarnate in his creation. But this takes us beyond the plan of phenomenology.

Now, within these limits, modest as they are, something very important seems to me to be furnished by experience in favour of the speculations of metaphysics.

On one side, when once we have admitted the above-mentioned transposition of the concept of consciousness, nothing any longer stops us from prolonging downwards towards the lower complexities under an invisible form the spectrum of the ' *within* '. In other words, the ' psychic ' shows itself subtending (at various degrees of concentration) the totality of the phenomenon.

On the other side, followed upward towards the very large complexes, the same ' psychic ' element from its first appearance in beings, manifests, in relation to its matrix of ' complexity ', a growing tendency to mastery and autonomy. At the origins of life, it would seem to have been the focus of arrangement (F 1) which, in each individual element, engenders and controls its related focus of consciousness (F 2). But, higher up, the equilibrium is reversed. Quite clearly, first from the ' individual threshold of reflection '—if not before—it is F 2 which begins to take charge (by ' invention ') of the progress of F 1. Then, higher still, that is to say at the approaches (conjectured) of collective reflection, we find F 2 apparently breaking away from its temporo-spatial frame to join up with the supreme and universal focus Omega. After emergence comes emersion. In the perspectives of cosmic involution, not only does consciousness become co-extensive with the universe, but the universe rests in equilibrium and consistency, in the form of thought, on a supreme pole of interiorisation.

What finer experimental basis could we have on which to found metaphysically the primacy of the spirit ?

*c.* Lastly, to put an end once and for all to the fears of ' pantheism ', constantly raised by certain upholders of traditional spirituality as regards evolution, how can we fail to see that, in the case of a *converging universe* such as I have delineated, far from being born from the fusion and confusion of the elemental centres it assembles, the universal centre of unification (precisely to fulfil its motive, collective and stabilising function) must be conceived as pre-existing and transcendent. A very real 'pantheism' if you like (in the etymological meaning of the word) but an absolutely legitimate pantheism—for if, in the last resort, the reflective centres of the world are effectively 'one with God', this state is obtained not by identification (God becoming all) but by the differentiating and communicating action of love (God all *in everyone*). And that is essentially orthodox and Christian.

APPENDIX

# SOME REMARKS ON THE PLACE AND PART OF EVIL IN A WORLD IN EVOLUTION

THROUGHOUT THE long discussions we have been through, one point may perhaps have intrigued or even shocked the reader. Nowhere, if I am not mistaken, have pain or wrong been spoken of. Does that mean that, from the point of view I have adopted, evil and its problem have faded away and no longer count in the structure of the world ? If that were so, the picture of the universe here presented might seem over-simplified or even faked.

My answer (or, if you like, my excuse) to this frequent reproach of naïve or exaggerated optimism is that, as my aim in this book has been limited to bringing out the *positive essence* of the biological process of hominisation, I have not (and this in the interests of clarity and simplicity) considered it necessary to provide the negative of the photograph. What good would it have done to have drawn attention to the shadows on the landscape, or to stress the depths of the abysses between the peaks ? Surely they were obvious enough. I have assumed that what I have omitted could nevertheless be seen. And it would be a complete misunderstanding to interpret the view here suggested as a sort of human idyll rather than as the cosmic drama that I have attempted to present.

True, evil has not hitherto been mentioned, at least explicitly. But on the other hand surely it inevitably seeps out through every nook and cranny, through every joint and sinew of the system in which I have taken my stand.

First : *evil of disorder and failure.* Right up to its reflective zones we have seen the world proceeding by means of groping and chance. Under this heading alone—even up to the human level on which chance is most controlled—how many failures have there been for one success, how many days of misery for one hour's joy, how many sins for a solitary saint ? To begin with we find physical lack-of-arrangement or derangement on the material level ; then suffering, which cuts into the sentient flesh ; then, on a still higher level, wickedness and the torture of spirit as it analyses itself and makes choices. Statistically, at every degree of evolution, we find evil always and everywhere, forming and reforming implacably in us and around us. *Necessarium est ut scandala eveniant.* This is relentlessly imposed by the play of large numbers at the heart of a multitude undergoing organisation.

Second : *evil of decomposition.* This is no more than a form of the foregoing, for sickness and corruption invariably result from some unhappy chance. It is an aggravated and doubly fatal form, it must be added, inasmuch as, with living creatures, death is the regular, indispensable condition of the replacement of one individual by another along a phyletic stem. Death—the essential lever in the mechanism and upsurge of life.

Third : *evil of solitude and anxiety.* This is the great anxiety (peculiar to man) of a consciousness wakening up to reflection in a dark universe in which light takes centuries and centuries to reach it—a universe we have not yet succeeded in understanding either in itself, or in its demands on us.

Lastly, the least tragic perhaps, because it exalts us, though none the less real : *the evil of growth,* by which is expressed in us, in the pangs of childbirth, the mysterious law which, from the humblest chemism to the highest syntheses of the spirit, makes all progress in the direction of increased unity express itself in terms of work and effort.

Indeed, if we regard the march of the world from this standpoint (i.e. not that of its progress but that of its risks and the efforts it requires) we soon see, under the veil of security and harmony which—viewed from on high—envelop the rise of man, a particular type of cosmos in which evil appears necessarily and as abundantly as you like in the course of evolution —not by accident (which would not much matter) but through the very structure of the system. A universe which is involuted and interiorised, but at the same time and by the same token a universe which labours, which sins, and which suffers. Arrangement and centration : a doubly conjugated operation which, like the scaling of a mountain or the conquest of the air, can only be effected objectively if it is rigorously paid for—for reasons and at charges which, if only we knew them, would enable us to penetrate the secret of the world around us.

Suffering and failure, tears and blood : so many by-products (often precious, moreover, and re-utilisable) begotten by the noosphere on its way. This, in final analysis is, what the spectacle of the world in movement reveals to our observation and reflection at the first stage. But is that really all ? Is there nothing else to see ? In other words, is it really sure that, for an eye trained and sensitised by light other than that of pure science, the quantity and the malice of evil *hic et nunc*, spread through the world, does not betray a certain *excess*, inexplicable to our reason, if to *the normal effect of evolution* is not added the *extraordinary effect* of some catastrophe or primordial deviation ?

On this question, in all loyalty, I do not feel I am in a position to take a stand : in any case, would this be the place to do so ? One point, however, seems clear to me, and it is sufficient for the moment as an orientation : that in this case (just as in that of the ' creation ' of the human soul—see note p. 169), complete liberty is not only conceded but offered by the phenomenon to theology, so that it may add precision and depth (should it wish to) to the findings and suggestions—always ambiguous beyond a certain point—furnished by experience.

In one manner or the other it still remains true that, even in the view of the mere biologist, the human epic resembles nothing so much as a way of the Cross.

*Rome, October 28, 1948*

# Index